A HUNDRED BILLION STARS

A HUNDRED BILLION STARS

· MARIO RIGUTTI

TRANSLATED BY
MIRELLA GIACCONI

The MIT Press
Cambridge, Massachusetts, and London, England

The MIT Press
Cambridge, Massachusetts, and London, England

This book was set in Palatino by The MIT Press Computergraphics Department and printed by Halliday Lithograph in the United States of America.

Library of Congress Cataloging in Publication Data

Rigutti, Mario.
   A hundred billion stars.

Translation of: Cento miliardi di stelle.
   Bibliography: p.
   Includes index.
   1. Stars—Popular works. I. Title.
QB801.6.R5313 1984   523.8   83–906
ISBN 0–262–18111–8

CONTENTS

·A FOREWORD THAT IS ALSO MEANT AS A JUSTIFICATION

The first time I wrote a book on astronomy for the general public was fifteen years ago. I put a lot of time and effort into it, and I did my best to make it easy to understand and pleasant to read. Although the book did not turn out too badly, it was not very successful, and a few years later the publisher decided to get rid of the unsold copies to make room for its latest publications. The book is out of print today, and while it may be flattering to be the author of a bibliographical rarity, the experience was neither rewarding nor encouraging. As a result, it remained my one and only contribution to popular science, and I forgot all about it.

Time has gone by and many things have changed. For one thing, much has happened in recent years to stir up a new interest in the sky—the flying-saucers craze, science fiction, the moon landings, space probes, Mars, the comet Kohoutek. All of a sudden astronomy has become quite popular, particularly with young people. With the enthusiasm typical of their age, these young amateurs have started building telescopes, studying maps of the sky, and making observations.

Unfortunately, this new interest has not found the proper response in the Italian press. On the one hand, there is a conspicuous lack of professional work in the popularization of astronomy. (Actually not too many professionals are fit for it, for one reason or another.) On the other hand, the field has become the happy hunting ground of writers who have little or no scientific background and whose approach to scientific writing can only be described as yellow journalism. The same goes for our self-proclaimed science correspondents and television producers. In the hands of these people, science, and astronomy in particular, has become a vehicle for cheap thrills, a way to fascinate the public, make it awestruck, overcome with the grandeur of it all. Their books (articles, television programs) are nothing but a rehash of the same old stuff spiced with grandiloquence and breathtaking pictures. More often than not, our publishers give a helping hand to the work of the popularizer, who is so generously giving of time and knowledge, by marketing it with a slogan sure to catch the reader's fancy. The book sells. It is even read. Most people do not learn very much because the book is badly written and superficial. It lingers on trivia and glosses over the difficulties (thereby ensuring that nobody will understand). And by using the wrong examples it generates a lot of confusion and misunderstandings. But one thing it never lacks is flights of fancy designed to impress the naive reader—which is all the book was meant to do. To the deep gratification of its author, such a work may even receive a prize.

In light of all this, writing this book was for me a matter of civic and cultural responsibility. I am not at all sure that I have succeeded in doing what I felt was almost a duty, but I could not keep on worrying, rewriting, and improving it forever. Setting aside my lingering doubts, I finished the work and sent it off to the publisher. Whether it will be of any help, I do not know, but I certainly hope so.

I have attempted to tell what is currently known about our galaxy as simply as I could, and I have confined myself to this part of the universe because one book is not nearly enough to deal properly with all of it. I have also tried to clarify some common misconceptions concerning the world of science, namely, the beliefs that the scientist is possessed of infinite knowledge and that astronomy is the least human of human activities. Let me make clear that it is not my intention to belittle science or to suggest that there is nothing to it and that everybody can understand everything. But I would be happy if in the end the book were to prove that science is the work of humans—not superhumans.

This book is addressed to the lay reader who is genuinely interested in astronomical research, its methods and findings. As taxpayers, moreover, such readers have the right to know whether the money the government takes away from their hard-earned wages is spent on enterprises they can be proud of, at least as members of the human community. I also hope that this book will appeal to young people because I have become convinced of the educational value of astronomy, in that it can lead the mind to the abstract by way of the practical, by attention to facts and their observation, by the development of the tools by which astronomy proceeds. Furthermore, I am mindful that while today's youth grows up very quickly in many respects, it is still young enough to be easily taken in by arrogant academics, glamorous reporters, and special correspondents as ignorant as they are presumptuous.

Deciding to write a book of this sort was easy, particularly since I seemed to be one of the few people in Italy who had not written one. But saying it is one thing, doing it another. My first problem was one of style, which is not a trivial problem. One can decide not to write in a stuffy academic style, but among the styles remaining it is not easy to hit on the right one. I finally decided to write in the same way that I speak to the young people it is my good fortune to meet as a university professor. This way I can be freer, more immediate and more genuine. This does not mean that I have only addressed myself to the young, but rather that I have imagined my readers to be younger than their years and happy to play with me, if only for a little while, and to return to the time when they still had the capacity to look around and wonder. At heart, we are all children before the universe that surrounds us. I do not know whether this solution to my stylistic problem is the right one. These things have a way of becoming obvious only afterward, when it is too late.

There is something else I would like to say, even at the cost of not being taken seriously, to make sure that the rules of the game are clear from the start. Although I like astronomy well enough to make it my life work, I honestly do not believe that it is the most beautiful of all sciences. But it does not make any sense to rank the sciences in order of beauty or interest. What is important is the research, the clarification of the obscure, the understanding. It is undeniable, however, that astronomy has a special fascination, compounded of magic and poetry, dark boundless spaces and glittering stars. Perhaps this is because it was the first science cultivated by humanity and was associated for so long with magic and religious rites. Yet while it is true that the universe is full of unanswered questions, the same can be said of the earth, the sea, and all living things. Look closely at a plant and observe the slow unfolding of its blossoms. And ask yourself How? and Why? Or take a drop of water and observe it under a microscope. More hows and whys to be answered. Or ask yourself what enables our brains to read a book and remember it. Or what enables our brains to ask how our brains work. I could go on forever listing all the things that are just as worthy of our attention as the sky, but the point I am trying to make is that if I write about astronomy, it is only because this is the one subject that I can talk about with some competence.

One more thing. There is an issue that troubles many scientists today, much more so than in the past. Science is the work of historical human beings, living in their own specific times. Consequently, it is conditioned by all of humanity's activities, interests, taboos, fetishes, illusions, and hopes, as well as their socioeconomic conditions. Doing science is not simply a process of learning how the world is made in total detachment from human concerns. Science is not a thing apart, and the scientist cannot think of living outside society, outside its miseries and fortunes. If in the past science developed in a certain way, it is because it was useful to somebody; otherwise it would not have been pursued. If today it continues to develop along the same lines, it is because, presumably, it continues to be useful to somebody. If science is useful, then scientists are useful; they are useful to those who benefit from their work. But the world is so regulated that, at least for the present, this work does not seem to benefit the laborer, the unemployed, the starving Indian or the African blinded by glaucoma, the illiterate, the very poor, the dispossessed. Most of humanity, I daresay, does not profit from their work. Should we still do science, then? And if the answer is in the affirmative, as I believe, how should it be done? I am afraid that the next generation will have a much more dramatic confrontation with this issue than ours ever did.

Finally, I wish to stress that this is not a textbook. Not only is it not going to explain everything in the smallest detail, but it has many gaps (the great planets, just to mention one). I trust the publisher will give it a friendly and easygoing look that will emphasize—even visually—the fact that science can be treated seriously, that is, correctly, without melodramatics or scholarly stuffiness. In sum, this book is in the nature of an encounter. The reader and I are going to walk together for a while and talk about things that interest us both. I will be the only one to speak, actually, but I promise to make my monologue as easy to take as possible.

Unless I am much mistaken, our quick foray into the world of the stars will raise many more questions than it answers. But if this book motivates readers to carry on the exploration on their own, I shall consider it to have been well worth the effort.

·A SHORT FOREWORD FOR THE AMERICAN EDITION

When I wrote this book it did not occur to me that some day it would be translated into English and published in countries that, unlike Italy, have long and well-established traditions in the popularization of science.

Italian scientists have always been somewhat disdainful of popular-science writing, which may seem surprising when you consider that Galileo wrote many of his books in Italian rather than Latin, the language of scholars, for the specific purpose of reaching a wider public. With Galileo pointing the way, Italy might have been expected to develop and nurture this kind of tradition, but unfortunately it did not happen. Of course, we can find other examples of popular-science writers in our past history—and some are truly outstanding—but they remain isolated cases. The academic mentality has always prevailed here, and the professor has always spoken *ex cathedra*. It has taken two world wars, profound changes in social structures and mores, and the widening of cultural horizons to crack science's ivory tower.

It is still only a crack, however. Although the situation has improved, it is far from satisfactory. Even today, popular-science writing consists essentially in listing and describing facts and phenomena without any serious attempt to make the reader understand and share in the process of discovery. To be understood, things must be explained, and this means that the writer must teach. Most of our books, instead, are filled with gratuitous statements that have to be taken on faith, and all too often the lack of substance is covered up with grandiloquence.

A book must be seen in the context in which it was produced. This one was written in Italy, where science writing for the general public is very poorly done, at least in my opinion, and fails in its most important task. Since I do not like this state of affairs, I decided to try a new, shall we say more democratic, approach. I must say that I do not seem to have had a large following so far. Perhaps the sun is too hot here and we are all a little lazy. Also, as you know, no one is a prophet in his own country.

These things needed to be said because certain aspects of the book may be puzzling to an American reader. The fact is, science reporting is much more professional in the United States. That is not to say that all the American books I have read are masterpieces. But there are certainly quite a few that I would be glad to call my own.

I am also aware that my style of writing may sometime appear a little paternalistic. If I sound like a wise man condescending to teach out of the goodness of his heart, rather than from a sense of duty, it is only because, not being an Erskine Caldwell or an Arthur Miller, I am not always able to express my thoughts with immediacy and warmth, however much I may wish to do so.

As I was reading the typescript of the translation for the American edition, I realized that some things have changed since 1978. For example, we now have a beautiful radar map of the surface of Venus, the number of known pulsars has risen to 350, and the fastest pulsar has a period of 1.558 milliseconds. I could—and perhaps I should— have revised the text, but I preferred to leave it as is because these findings, though important, do not change the overall picture I have tried to present.

I feel I owe a lot to the translator, and I wish to thank her for doing the work, not only with competence, but with care and sensitivity. Naturally, I am not the best judge of it because English is not my mother tongue, but if I had written the book in English, this is the way I would have wished it to be.

February 1983
Naples

· ACKNOWLEDGMENTS

It is difficult to write about science for people who do not "do" science, and it would not surprise me if this book did not wholly satisfy everybody; there are just too many difficulties to overcome in the writing of a good popular-science book. On the other hand, while it is clear what such a book should not be, it is not at all clear what it should be.

In any event, I would like to thank a few people for what little good I have been able to accomplish: my wife, first of all, who has helped me in many ways, besides accepting graciously the further curtailment of my free time—I owe her very much, as usual; M. A. Santaniello, who read and criticized the text, enthusiastically sharing with me her brief but intense experience as a high school teacher; my friend Ronald R. La Count of the National Science Foundation, who patiently provided me with many of the photographs that illustrate the text; and finally, all the colleagues and friends whose help, so generously given in many ways, has enabled me to write a much better book than I could have written by myself.

Plate 1
The solar corona, photo-
graphed by R. Ciappi of the
Italian group of the Capodimonte Astronomical
Observatory at Atar, Mauri-
tania, during the total solar
eclipse of June 30, 1973.

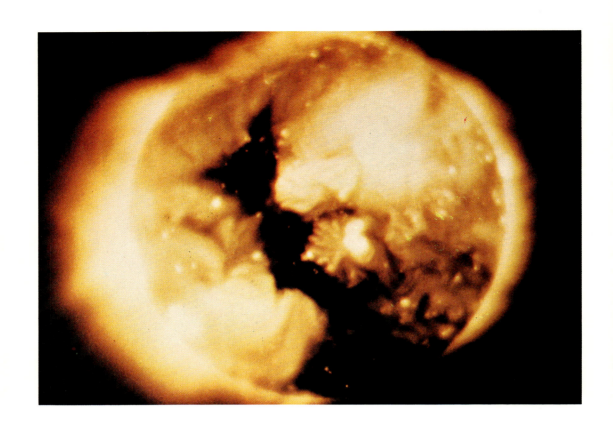

Plate 2
Photograph by *Skylab* of
the sun's radiation in the
wavelength intervals from
2 to 32 and from 44 to
54 Å. The radiation emit-
ted—in this case, x rays—
originates in the corona
(temperature between 1 and
2 million degrees kelvin; see
note 12). The photograph
shows large-scale struc-
tures, a "coronal hole," and
other regions of low
emissivity.

Plate 3
Photograph of Venus taken
by *Mariner 10* in ultraviolet.
The thick cloud cover is
clearly visible.

Plate 4
Photograph of the Martian
soil taken on July 21, 1976,
by *Viking 1* at about noon
(local time). The reddish
color of the soil is due to
iron oxide. Dust suspended
in the atmosphere is re-
sponsible for the atmo-
sphere's rosy tinge.

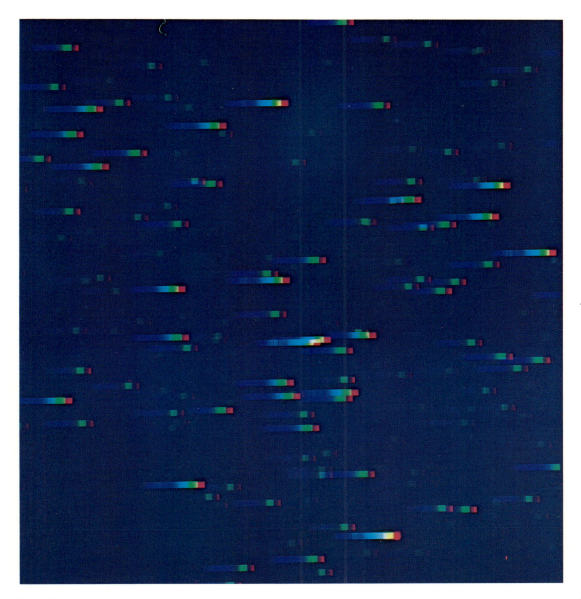

Plate 5
Spectra of the stars in the
cluster known as the
Hyades.

Plate 6
Circumpolar stars of the Southern Hemisphere. The faint nebulosity on the right of the South Celestial Pole is the Small Magellanic Cloud, and the nebulosity above the pole is the Large Magellanic Cloud, which are two small galaxies, satellites of our own. The white star whose track appears on the right is Achernar, and the white star just risen at the left is Acrux. The photograph was obtained with an exposure time of 24 minutes.

Plate 7
The Lagoon nebula in the constellation of Sagittarius.

Plate 8
The Trifid nebula in Sagittarius. Observe the different
colors of the stars.

Plate 9
The Omega nebula in
Sagittarius.

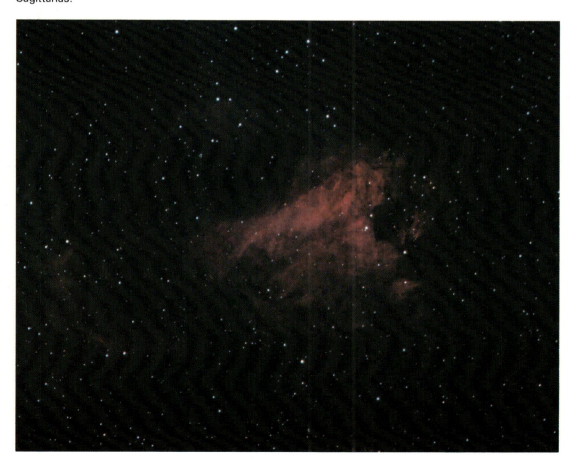

Plate 10
The Eta Carinae nebula in
Carina, a constellation in the
southern hemisphere.

Plate 11
The Aquilla nebula in the
constellation of Serpens.

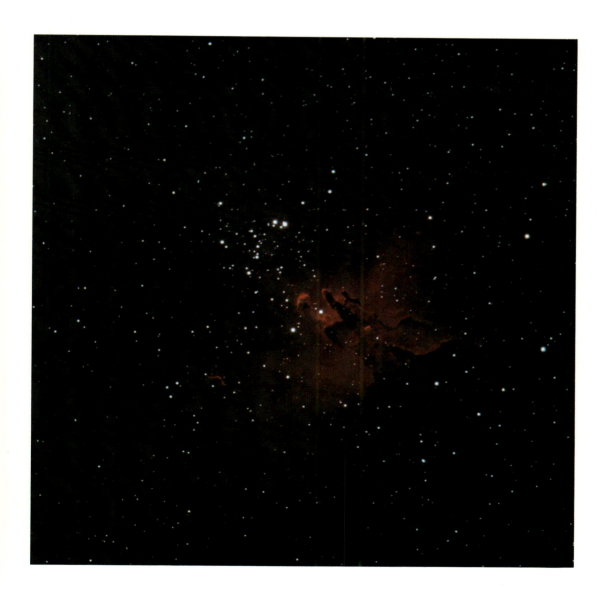

Plate 12
The nebula in Orion. Different colors correspond to
different temperatures.

PART I
THE EARTH'S
NEIGHBORHOOD

• The Eclipse

This eclipse the scientists make
is really something special, I can't deny it.
But every time it's the same story—
we end up by getting gypped.

Last year, just to see it,
my brother came especially from Frascati,
stood an hour with his eyes open wide,
and couldn't see a single star.

If the sky is always cloudy,
next time they make it
nobody will believe them.

And that would serve them right.
If it's cloudy, who can see it?
Why do they tell us? Why do they make it?

Trilussa

· PROLOGUE

In an interview Frank Bormann remarked, "What I would like to say is that it was not just a question of looking at the earth, but of looking at it within the universe. Sure it was the earth, our earth, but now it was a little smaller. . . . The earth, seen from lunar distances, is so beautiful, so quiet, so peaceful that if we had not known its problems, to tell you the truth, we would have believed it a small, silent, peaceful, and lonely world."

But the truth is, the earth is neither peaceful nor silent. As we all know, including Mr. Bormann, the blue world he saw from the moon is going through what is perhaps the most dangerous period of its existence. An animal species that has multiplied into billions of individuals may be about to destroy itself and all the living things nature has created over millions of years. People on earth are hungry, thirsty, and afraid. Actually, not everybody is hungry. Some (like us, for example) have so much stuff to eat that they throw it away. And not everybody is thirsty. Some (like us, for example) have more than enough water for every conceivable use. Yet when we go hiking on the mountains, where water is as clean and pure as a child's smile, we give up its cool blessing for a silly drink whose sparkle is due to the trick of adding a little carbon dioxide. Almost everybody is afraid. Some (like us again) live in fear as if in a new dimension—total fear, without hope. For violence, poverty, disease, intimidation, exploitation, cowardice, cynicism, and war are the threads that form the fabric of today's world.

Yes, the earth is a small and paltry thing. We knew it already and we did not need to look at it from a spaceship to realize it. In any case, so few people have seen it that way that it makes no difference. So the earth is small. All right. But since we are very small, to us it looks large—it *is* large—and we like it (although we might wish to change it), and we resent having its smallness emphasized, as if this fact could ever lighten the burden of responsibility for having made such a mess of it. For even a small mess is still a mess. No venture into space, presumably inspired by the peaceful desire for scientific knowledge, can change this fact. There is no war in the world today in which the sons of Western civilization are not involved in some way, at the very least as arms merchants. We have become so insensitive and cynical that we listen with the same indifference to a peanut butter commercial and to the news of a massacre in one of the many wars being fought around the world. For radio and television, peanut butter and dead people are on the same level. And for us as well. But we make quite sure that our space travelers do not contaminate the lunar environment with foreign microbes. I do not mean to imply that precautions were superfluous. In a certain frame of mind every one of us must agree that they made very good sense. What is frightening is seeing the two sides of people or, worse, the thought that perhaps there is only one side, that people indifferently love or destroy microbes or other people depending on what is best for them. All of which fits in perfectly with the general madness of our times.

It is all so hopeless that you feel like giving up and consigning the whole world to the devil. You feel that way, but then you discover things in yourself—memories, feelings, a faith in humanity that by now may be more of a vice than a virtue—that prevent you from giving up and push you to keep working in the absurd hope of doing something that will change reality for the better, of contributing in some small measure to the betterment of humanity, society, and the world we live in—a world that could be so unbelievably beautiful!

We must learn to work for people, not for progress, or science, or scholarship, or art, or anything else, but for people: in their youth, in their maturity, and in their old age.

You will ask, What has this to do with anything? Maybe nothing. But you know how it is. One starts talking about one thing and ends up talking about something else. I started by reporting astronaut Bormann's impressions of the earth from far away, which seemed like a good beginning for a chapter devoted to the earth's neighborhood, and promptly slipped into a nonastronomical matter.

I could easily slip into other nonastronomical matters. For example, I have every right—and probably the duty—to discuss the overly technical and specialized slant we have learned to give to our activities, as if each of them could be carried out without taking the others into account. Part of the moral and intellectual confusion of our society may indeed be due to the kind of order that gives to each thing its proper time and place. Whom is this order good for? I recall that when I was a boy there were signs on the walls of offices and factories that said, "One does not talk politics here. One works." I cannot deny that there was some good in that sentence—when you decide to work, you have to keep your mind on it—but the fact is that politics could not be talked about anywhere else, and soon people lost the will to do so. Coming back to our case, this habit of talking about things "in the proper place" has the effect that no problem can ever be discussed in its entirety and complexity, and in the end every problem is distorted by focusing on this or that aspect of it. For example, to stick to the field of science, should astronomers interested in space research worry about the social relevance of their work? Should researchers worry whether the money they are about to spend, which is public money, is well spent? One might say that it is not up to the scientists to make certain decisions, that responsibility for this type of choice rests with other people, and that if money is given to them, it is because it has been decided—in the proper place—that it is money well spent. But I might argue that it is not right to keep passing the buck. If everything we do is done for humanity (and for whom else should we do it?), should not each of us be aware of the significance of our actions and of the effect that may have on people's lives?

Perhaps, then, what I was saying at the beginning of this chapter did have some meaning and did have a place in this book. I will not delete it. And now, back to the earth and its environs.

· A FEW NUMBERS

I shall not burden my wise reader with silly calculations proving that space travel is not possible by bicycle, train, racing car, plane, or rocket. We all know that it would take too long. Even at a constant speed of 1 million km/hour[1] (at this fantastic speed we would get to the moon in a little over 20 minutes!), it would take us 9,000 years to reach the star nearest to the sun and come back. (What was happening on the earth 9,000 years ago? Humanity was living through its Mesolithic period. What will be happening 9,000 years from now?) The distances are just too great. Even a round trip to Pluto at that speed would take as long as 1.5 years. And to think that Pluto, the outermost planet in the solar system, far as it is, relatively speaking is just around the corner.

Obviously, it does not make any sense to talk about space travel or to think that we can conquer the universe. The fact that we have gone to the moon does not mean a thing. We are in the same position as the man who has just learned to swim and, having ventured a bit into the water, starts dreaming of the day he will be swimming around the world. Let's face it: We are

prisoners of the earth or, at best, of its immediate neighborhood.

We could imagine traveling on a beam of light, which means moving at a speed of 1 billion km/hour (1,000 times faster than before!). For a while it seems to work very well; in 6 hours we are on Pluto, and in 4.5 years we arrive at the nearest star. But to travel across our galaxy, from one end to the other, would take 100,000 years. And it would take 2 million years to reach the nearest galaxy. And it would take. . . . Enough of this. I think I have made my point.

Since models sometimes help to visualize things, let us make a model of the solar system. Let the sun be a ball 10 cm in diameter; then Mercury and Venus are two tiny spheres respectively 0.3 and 0.9 mm in diameter located 4 and 7 m from the sun. The earth is a sphere 0.9 mm in diameter that revolves around the sun at a distance of 10 m. Mars is about 15 m from the sun and has a diameter of 0.5 mm. Next come the large planets, the important members of the family: Jupiter (10 mm in diameter), Saturn (9 mm), Uranus, and Neptune (each 3.5 mm). They revolve around the sun at distances of about, respectively, 50, 100, 200, and 300 m. Pluto, the farthest planet, has a diameter of about 0.5 mm and its orbit takes it as far as 500 m from the sun.

Our solar system is truly insignificant. If the small spheres were all in a line and revolved around the sun all together as if mounted on the spoke of a wheel with the sun at the hub, perhaps they could be noticed. But since the planets move in their orbits at different speeds,[2] at any given moment, practically speaking, they appear scattered here and there at random. When Neptune and Pluto are on opposite sides of the sun, for example, they are 800 m apart—quite a distance for two tiny spheres 3.5 and 0.5 mm in diameter. Seeing them is short of miraculous.

This is our first discovery. Our solar system is scattered far and wide, and if we could put it all in a box (we could try it with our model), the box would be practically empty.

This observation deserves further attention. If we measure the masses of the planets (see table 1), we find that their sum is about 2.7×10^{27} kg.[3] (The mass of the sun is about 2×10^{30} kg, or 740 times that of all the planets put together.) If we distribute this planetary mass uniformly throughout the space occupied by the solar system, that is, in the sphere whose radius equals the distance from the sun to Pluto, and hence whose volume is about 860×10^{27} km^3, it turns out that the average density in the sphere is

The earth seen from the moon. Photograph taken during the *Apollo 17* mission (December 7–19, 1972).

approximately 3 g/km³, which is 3 million-millionths of a gram per cubic decimeter ($0.000000000003 = 3 \times 10^{-12}$ g/dm³), which is nothing, or almost nothing.[4]

Our solar system is almost nonexistent. What exists is the space it defines, just as a space exists after you build something—a wall, an arch—where before there was nothing. In our case somebody (?) scattered a sun-ball and various planet-balls where there was nothing, thereby creating the space of the solar system.

•DIGRESSION

In astronomy, as you may have noticed, as soon as you stick your head out the window to have a look around you run into numbers that make your head spin (the so-called astronomical numbers). There are writers and lecturers who exploit this fact to arouse in their audiences feelings of wonderment or awe, fear or enthusiasm, and to surround their tired old words with a mystical halo. Much like the sorcerer's fancy dress and mysterious rituals, these conjuring tricks are used by people to give the impression that they are so involved in such deep things that they are worthy of our reverence, that unlike the rest of us, they live in a state of constant detachment from the banalities of everyday life. If any of them are sincere, they must be maniacs because real people simply cannot spend all their time contemplating the majesty of the universe. Even the greatest astronomer finds the time to take a shower, eat a steak, listen to music, go on a picnic, get mad, buy a car, make love, worry about inflation, discuss politics, and play. The fact is, no matter how passionately one loves astronomy, very little of this passion is left after working hours, even though the term "working hours" must be taken here with a certain latitude.

All of this is pretty obvious. People cannot spend their entire lives stretched to the breaking point, breathlessly engaged in scrutinizing the depths of the cosmos, unveiling the mysteries, wondering at the wonders. Yet because of certain historical and social conditionings and because of a number of prejudices that scientists (and particularly astronomers, even if mediocre—and especially if mediocre) cannot shed, popular-astronomy literature is full of exclamation points, high-flown bombastic language, comparisons between the smallness of the earth and the magnificence of creation, elevated but trite thoughts, abyssal depths where the spirit falters, rosy dawns and sunsets of poignant beauty, and, for contrast, the chirping of tender birds (the nightingale is everybody's favorite), a flight of butterflies, the slow unfolding of flower buds, and celestial symphonies of harps and violins.

Let us be done with it. One can understand a show of sentiment, an occasional flight of fancy, a touch of fantasy or emotion, but it is unforgivable that such things should be used to bamboozle readers and to drag them oohing and aahing through the world of astronomy as if it were a museum of wonders.

The task of scientists who share their experience, of professors who teach, or writers who popularize, is not to impress their audience or to create a mystical sense of the world, but rather— and this is particularly important when addressing the young— to widen the horizons of knowledge and to teach how to look at things and phenomena with an alert eye, a critical spirit, and a clear unbiased mind. If any emotion must be aroused, it should not be a feeling of awe before a world too big to be conquered, but rather the joy that comes from understanding. It does not matter that our understanding is limited and that science is full of shadows and dark corners. What is important is to communicate, along with a few facts, humanity's ability to emerge gradually through constant effort and the work of generations from ignorance, irrationality, and superstition. We do not know everything. On the contrary, strange as it may seem, we know hardly anything, but it is also true that as time goes by we learn a little more. That is all that counts.

The world of celestial objects raises countless questions, and these constitute the subject of astronomy. Astronomers, physicists chemists, mathematicians, engineers, technicians, biologists, and philosophers, in their own ways, with their different qualifications, cooperate in the effort of answering these questions. Every answer usually raises more questions, but this is not, except in appearance, an increase in ignorance. Thus people who popularize scientific research have the duty not to cheat the reader by draping their knowledge in a mantle of magic. Magic may be good for them, and perhaps for their colleagues, but certainly not for education and the development of a critical sense.

For all these reasons and more, I will do my best not to weave a magic spell around the minds of the readers. If there is a poetry of science, they will have to discover it by themselves.

·THE OBJECTS OF THE SOLAR SYSTEM

Let us get on with the business. We were talking about the space defined by the solar system. There are many things in it: the sun, first and foremost; then the 9 planets and their satellites, 33 in all; then, in a belt between Mars and Jupiter, some 100,000 asteroids;[5] and then comets, as many as you want—50,000, 100,000, 200,000. In addition, there are meteorites and micrometeorites beyond count. On its way through space the earth collects almost a billion of them a day, for a total of about 5 tons. Since the meteorites travel at about 50 km/sec and the earth at about 30 km/sec (although nobody notices), these objects enter the earth's atmosphere fast enough (30 to 70 km/sec) to burn up by attrition with the air and dissolve into gas. Almost everyone has seen one of these flashing apparitions we call shooting stars. It is only the larger meteorites that come through the atmosphere without vaporizing completely and reach the ground. They number less than 10 a day and come in all sizes. Some of the largest ones—perhaps asteroids captured by the earth's gravitational field—can be frighteningly large. You need only look at the meteor crater in Arizona (figure 1) to have an idea of what the earth can encounter.

In the solar system there is also a large amount of dust. Have you ever heard of the zodiacal light? If you have not, I shall tell you that it is sunlight scattered by interplanetary dust, or, more precisely, by the particularly dense dust on the ecliptic plane, which, as you probably know, is the plane in which the earth's orbit lies.

The ecliptic plane and the plane of our horizon can never coincide or be parallel (except at points located on the Arctic and Antarctic circles); hence the ecliptic plane is more or less tilted on the horizon depending on the latitude of the observer and the time of observation. It is not difficult to locate. The orbital planes of the planets and moon are nearly coincident with the ecliptic plane; thus to draw it roughly on the celestial sphere, all you have to do is join the visible planets and the moon with an arc of circle. Shortly after sunset or before sunrise, provided the sky is quite dark and the atmosphere transparent, you may be able to see along the ecliptic a diffuse glow in the shape of a cone whose base is on the horizon, where the sun has set (or where it will rise), and whose vertex is in the sky. This is the zodiacal light, and, as I said, it is sunlight scattered by countless specks of dust floating in space. We see them just as we see all those

Figure 1
Meteor Crater in Arizona,
1,300 m across and 180 m
deep. The edge of the crater rises 45 m above the
surrounding land.

luminous points in a dark room crossed by a beam of light. The reason why we see the beam of light is that each speck of dust in the beam's path scatters the light; otherwise we would not see it.

Interplanetary space is also crossed by swarms of particles. Some swarms consist of nuclei of hydrogen and helium as well as nuclei of heavier atoms; these are called cosmic rays because they come from everywhere, from the cosmos, by day and by night. Other swarms consist of particles of the same type as cosmic rays but not as fast, and hence less "energetic," which come from the sun as a result of short-lived, explosive events occurring in the upper layers of the sun.

Finally, interplanetary space is filled with a flux of hydrogen nuclei, electrons, and nuclei of heavier atoms that the sun emits without stop as if it were evaporating. It is the so-called solar wind. Today we can measure this pervasive flux of solar particles with instruments on board the artificial satellites, but in a way we can also see it by the effects it produces. If it is, as its name would have us believe, a wind, then it must cause something to blow away. If you look at a comet's tail (figures 2 and 3), you will see that it is always on the side of the comet opposite to the sun; hence, contrary to what one might expect, every comet moves away from the sun tail first. Almost all comets come from very far away, and after passing quickly around the sun, they recede into space, often not to be seen again. When it is still far away a comet looks like any one of the many objects that populate space, but in the vicinity of the sun it is an entirely different thing. The flux of solar particles and radiation cause transformations in its structure that result in the emission from it of gas, which the unceasing solar wind blows away from the sun. This is the tail— a stream of gas and dust, sometimes hundreds of millions of kilometers long, that keeps on dissolving into space and renewing itself as long as the disintegrating force emanating from the sun persists. If the comet travels in an elliptical orbit, like the planets, it will come back—perhaps in a few years, perhaps not for a long time—but it will come back. And then the sun will consume a little more of it, and a little more at its next return, and so on until one day the sun will dissolve it completely. What will then be left will be gas and dust (and if the comet has exploded, stones and pebbles as well) scattered along an elliptical band, a memento of what was once a comet.

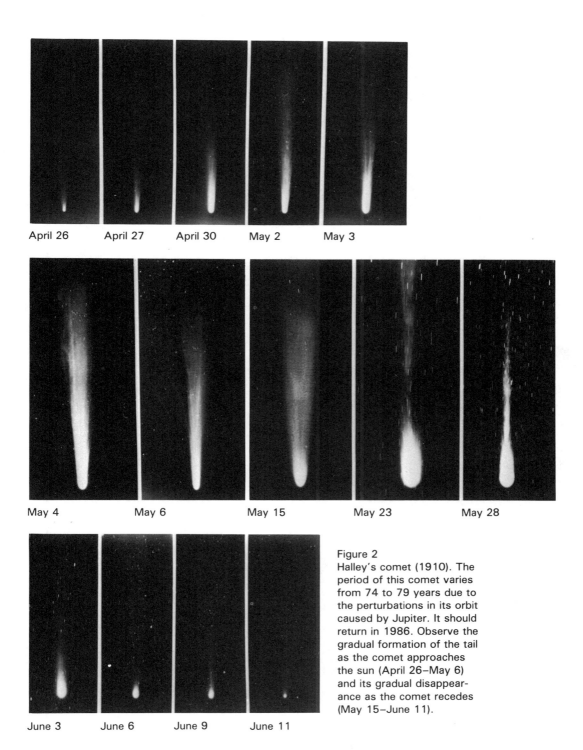

April 26   April 27   April 30   May 2   May 3

May 4   May 6   May 15   May 23   May 28

June 3   June 6   June 9   June 11

Figure 2
Halley's comet (1910). The period of this comet varies from 74 to 79 years due to the perturbations in its orbit caused by Jupiter. It should return in 1986. Observe the gradual formation of the tail as the comet approaches the sun (April 26–May 6) and its gradual disappearance as the comet recedes (May 15–June 11).

Figure 3
Comet Bennet (1970).

It is now time to take a look, only a quick look, at the various objects of the solar system and at the secret bonds that tie them to the sun.

• THE SUN—BIG BUT ALSO SMALL

Let us start from the sun, lord and master of the system. Of all the billions of billions of stars that exist, it is the only one whose surface we can observe in some detail. It is a very large object. Even if we confine ourselves to the ball that can be seen with the naked eye (disregarding the corona, that is, which normally cannot be seen), it is a sphere about 1.5 million km in diameter and with a mass of 2 thousand billion billion billion kg.[6] Quite understandably, the mind cannot grasp such a number—it is just too big and beyond our direct experience—but nothing can be done about that except to say that our sun is truly a very large object.

At this point you might say that you do not believe it or that you would like to know how we can measure the size and mass of an object so far from the earth (about 150 million km). Later on I shall have the opportunity to discuss the methods we use to determine stellar distances and other pertinent facts, masses included. You must also understand, however, that I shall not be able to justify every statement or to explain every last thing. Either I write a book that never ends, with little hope that anybody will read it, or I write a book of manageable size that will discuss a few of our findings and problems without frightening anybody away. Since I have taken the latter course—by choice, not out of laziness—the reader must trust me to some extent or, if not quite satisfied, let some of my assertions (and there will be many) be a spur to further reading.

The sun, then, is very big. But big and small do not have absolute meanings. In everyday language and when precision is not important, we can speak of big and small without quibbling. But as soon as you think about it, it is obvious that when we call something big (or small), we have something else in mind with which we are comparing it. If you ask for 2 meters of fabric and pay for 2 meters, and the storekeeper gives you only 1 meter, you would certainly say that he is dishonest or exceptionally absentminded because making a mistake of 1 meter out of 2 is a big mistake (either that or a big ripoff). But if you walk 50,001 meters, you can forget that last meter and say that you have been on a 50-

kilometer hike without being too far wrong. A meter, then, is big or small depending on what you compare it with.

The same is true for the sun. It is big if we compare it with the planets. It is immense if we compare it with our small selves. (But are we really that small? What if we compare ourselves with an apple seed? Or a grain of sand? Or a microbe?) Yet it is a smallish star because there are others that are much, much bigger. Note that this does not change in any way our subordinate position in relation to the sun.

·SOLAR ENERGY

Geological evidence tells us that hundreds of millions of years ago the temperature on earth was about the same as it is today. This means that the sun, which is practically the earth's only source of heat, was shining millions of years ago as it does today. But shining means emitting energy, and to emit energy it is first necessary to produce it. We are talking here of an enormous[7] amount of energy. I could jot down a number and we would have the usual "billions of billions" (about 400 thousand billion billion kilowatts; 1 watt $= 10^7$ erg/sec—see note 8 and the appendix) that does not mean anything. But perhaps you can get an idea of how much energy the sun radiates into space by consuming itself if you consider that all the winds on earth (breezes, storms, and hurricanes), the waves of the sea, and the production of clouds are the work of the sun. When you go swimming in the summer and the sea is as warm as a bathtub, think of how much it takes to warm up a pot of water, and then consider the fact that all the water that makes up the sea has been heated by the sun. Think, too, of the energy that we all expend to move, to speak, to work, or to fight and that is generated by the food we eat, which in turn is made up of plants and animals that grew because of the energy coming from the sun. Think of the energy that moves planes, ships, and cars. Isn't it fantastic how many things happen solely because there is a sun that consumes itself?

And there is something else to consider. Obviously, there is no reason why the sun should send energy toward the earth and not in all the other directions in space. If you agree with that, you must also agree that on every "point" as large as the earth and situated on the surface of a sphere having the sun at its center and whose radius is the distance from the sun to the earth, as much energy must fall as falls on the earth. Since the earth is

12,000 km in diameter and the earth-sun distance is 150 million km, it follows that the sun radiates into space 2.5 billion times the energy that reaches the earth. And this has been happening day after day, year after year, for many millions of years. Our sun is an infernal machine! Where does it get all that energy? And how long can it go on like that?

I shall not use farfetched examples and analogies, as popularizers usually do, to describe the type of calculation made by nineteenth-century scientists to explain this startling phenomenon. I will only say that all their efforts produced the same result. No matter what they thought, no matter which mechanism of energy production they invoked (and the possibilities were not many anyway), they were led to the conclusion that in no way could the sun have produced energy for more than 10 million years, which is much too short a time, considering that mammals had already appeared on earth about 200 million years ago. There was no way out of the impasse, and the sources of solar (and stellar) energy remained unknown, until the turn of the century, when physics was revolutionized. It took the combined efforts of the many physicists who gave us an insight into the world of atoms to understand how a star works, just as it took the knowledge (albeit not complete, not yet complete) of the world of cells and molecules to understand how living organisms work. It is not at all strange that it should be that way. Every natural phenomenon—and a star is also a natural phenomenon—is always the result of a complex of natural phenomena, and the key to understanding these complex natural phenomena is the ability to identify the elementary phenomena, the elementary facts, and the relations between them that produce complex phenomena. It is only after identifying the elementary facts and laws that we can arrive at the synthesis that clarifies what was once unexplainable.

Another short digression is needed here to avoid misunderstandings on a very important point. Saying that a once unexplainable phenomenon has become clear only means that at a certain time an explanation has been given for it that we find satisfactory. It does not mean that it is the only possible solution or that a solution once found will be valid forever. I could use many examples from the history of astronomy, or any other science, to illustrate this point, but I shall not mention them because it would take us too far out of our way. The history of any science is not simply a mosaic that little by little gets completed as new pieces are added in. The reason is that *doing* science is not simply a matter of looking at the world and recording how it is made, although this may be the *object* of science. When doing science, researchers are no mere spectators, but rather active participants conditioned by their own experience and preconceptions, which, by influencing the "questions" asked of nature, also influence the "answers," that is, scientific advances. This is why science, as a human activity, is also a product of its time and social conditions. Well aware of this fact, scientists no longer seek the Truth. Their task is to observe facts and phenomena and then to interpret them and to tie them together with a theory founded on a hypothesis, or a set of hypotheses, that stems, not from observed facts, but from the brains of historical human beings. Theory, in turn, suggests additional facts that need to be verified. Hence science is a dynamic process of adapting to reality a set of relations that transcend individual phenomena. Indeed, the purpose of science can be defined as the formulation of a set of relations that explain the known phenomena and permit the prediction of others.

Once in a while it happens that a theory is invalidated by the realization that certain pieces that appeared to fit well together do not allow others to fit in. As a result, progress comes to a halt. These are times of crisis and revision, during which the scientists strive to find a new design, by more or less drastic modifications of the old, that will allow them to fit together all the pieces—the old pieces that fit already (though they may have to be fitted in new ways) as well as the new pieces that did not fit before. Sometimes the modifications are so radical that the new design hardly resembles the old. Physics has gone through a number of these periods of transition and change, and some of them have altered the scientific edifice to its very foundations. Does all this mean that there is no guarantee that what we say corresponds to reality? This is probably a meaningless question. Reality is a complex of phenomena. Science creates a model of reality based on a set of relations that, as we said, proceed from assumptions born in our brains and elaborated by them. What science creates, therefore, is not an approximation of reality; nor does science identify the essence of reality. Rather, science creates a *model* of reality, a mental construct that arranges things within a logical-deductive system and allows predictions to be made. A scientific result, therefore, is the more satisfying, the more phenomena it can gather into its scheme and the more predictions (verified by experience) it can make. In this sense, the science of times past is as much science as ours. There is no other way to define science, and perhaps it does not make sense to ask whether the reality we reason about exists and has any meaning (what *we* mean by meaning) outside of our thoughts and cognitive acts. As soon as we ask such a question, we have already begun to reason; hence knowledge, whatever its meaning, cannot be detached from the person who wants to know.

I should also add that the physicists, or naturalists, who set out to give an explanation of the world are often led to believe that in the process they are really getting closer to understanding how the world of things would work even if human beings were not there; in any case, they dutifully introduce their interpretations of facts and descriptions of phenomena with the words "It all happens as if. . . ." In these five words, besides great intellectual honesty, there is also some of the sadness that comes from realizing the limitations of science done by human beings.

Let us return to the sun.

As I was saying, the physical phenomena that could not be explained within what today is known as classical physics led to a reformulation of many parts of physics and, in particular, to new views and new theories about the structure of atoms. On the basis of these new theories, new attempts were made to find a solution to the problem of the source of stellar energy. We shall discuss this point at some length in the chapter on stellar evolution, but for now let us see what are today's main ideas on the subject.

According to current theories (which should not change in any substantial manner for many years to come), the central regions of the sun have a temperature of approximately 15 million degrees and a pressure of about 300 billion atmospheres. Under these conditions fusion of the hydrogen atoms in the sun's core can occure naturally. This means that a process takes place whereby 4 atoms of hydrogen (the simplest and most abundant element in the universe) disappear to produce 1 atom of helium. As it happens, the mass of the helium atom is a little smaller than the sum of the masses of 4 hydrogen atoms; the extra mass has been completely transformed into radiating energy. More precisely, for every 1 g of hydrogen that turns into helium, 7 mg turn into energy. How much energy is that? A lot. Something on the order of 6.4×10^{18} ergs.[8] At home, where nobody is particularly careful, we use from 7,000 to 8,000 kilowatt-hours of electric energy per year on space heaters, water heaters, dishwashers, washing machines, television sets, record players, and all the other electric appliances found in most houses. Adding some gas, let us say that we use 10,00 kilowatt-hours/year, or 3.6×10^{17} ergs. It is rather remarkable to think that the energy liberated by the annihilation of 7 mg of hydrogen would be enough to satisfy all, or almost all, of my family's energy needs for 20 years, or, to put it another way, that the yearly energy needs of the 10 million families that live in Italy could be met by the energy obtained from the annihilation of 3.5 kg of hydrogen (or from the transformation of 500 kg of hydrogen into helium).

Since the mass of the sun's core is about one-tenth the solar mass, and since we know the latter, we can say that at the beginning of its existence the sun had at least 2,000 billion billion billion kg of hydrogen that could be turned into helium and radiating energy. Knowing how much energy the sun radiates into space every second and making a few calculations, we find that by consuming just the hydrogen in its core the sun could have emitted energy at the present rate for at least 10 billion years, give or take a million.

Enough astronomical numbers for the moment. But it is clear that we have found the mechanism, the source of stellar energy.

The rest of the sun enveloping the solar core is a gaseous mass, which blocks the hard and penetrating radiation (γ rays) produced in the core, absorbs it, transforms it, and reemits it as different (increasingly less hard) radiation. This flux of radiation travels through the solar body and in the end emerges and propagates into space. Enough of it reaches the earth to tell us the history of the sun and to give us life.

· THE SOLAR CORONA

The sun does not end where it seems to end when we look at it with the naked eye. We have many instruments that enable us to see that the gas sphere does not end in a clear-cut way, but goes on as an increasingly more tenuous nebulosity—the corona (plate 1)—that extends out to very great distances, beyond Mercury's orbit, Venus's orbit, and as far as the earth (so that one could say that we are actually inside the sun).

We do not really need instruments to see that the sun extends far beyond the limits we are apt to draw at first sight. We can see it with our own eyes during a total eclipse. Due to a set of circumstances (the sizes of the sun and the moon and their distance from the earth), total solar eclipses occur when the moon is between the earth and the sun in our line of sight. Under these conditions the moon blocks out the sun completely, enabling us to see its faint external regions extending out into space. With the naked eye, actually, the corona does not appear all the large—just 2 or 3 solar radii from the sun's edge (but that already means 2 million km). This is because, first, the corona quickly becomes much less brilliant as the distance from the sun increases and, second, the sky, although dark, is never quite as dark as the night sky.

You may wonder why you cannot obtain a total eclipse by placing something—a finger, a coin, or a screen—between the sun and your eye (or instrument) that will block out the sun's rays. You can indeed block out the sun with some kind of screen, but you cannot darken the sky (the earth's atmosphere), which remains lit up by the sun. Even close to the sun, where it is brightest, the corona is 100 to 1,000 times less bright than the sky (figure 4). Normally, therefore, it is invisible. But when the moon, which is outside the atmosphere, passes between the sun and the earth, it prevents the sun's rays from reaching the earth's atmosphere; thus the sky of the observer is no longer illuminated, and the corona becomes visible.

Once we have learned that the sun does not end where it seems to end, we naturally want to find out where it does end, that is, where the corona merges with the surrounding medium. (Needless to say, the density of coronal particles, essentially electrons, decreases with the distance from the sun.) This entails measuring

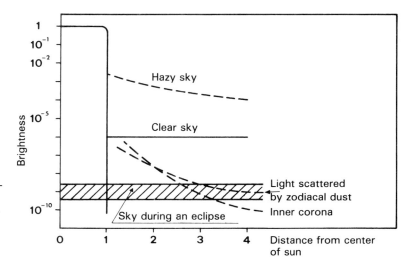

Figure 4
Plots of brightness versus distance from the sun's center of the sun, the sky, the inner corona, and the zodiacal light. 0 indicates the center of the solar disk; 1, the edge of the sun.

the density of particles in interplanetary space, and to do it we need instruments and techniques that have practically nothing to do with traditional astronomy.

We have to use space probes, rockets, satellites, and special detectors, and we end up doing a science so different in methods and techniques from the one to which we are accustomed, that it no longer even seems to be astronomy (plate 2). It has been called space physics, but it is only a new name for old problems and is used—rather gratuitously at that—to distinguish the astronomy that is concerned with celestial objects from the astronomy that deals with whatever else is in space. Coming back to the corona, measurements taken with the most disparate techniques have shown that the density of coronal electrons (I am not going to explain what electrons are; what you know about them is certainly enough for our purpose), that is, the number of coronal electrons per cubic centimeter, is 9 billion near the edge of the sun, 3 billion at a distance of 1 solar radius from the edge, 80,000 at 5 radii, 13,000 at 10 radii, 1,600 at 20 radii, 160 at 50 radii, 30 at 100 radii, and 5 (about 100 times more than in interstellar space) at 215 radii, that is, near the earth. Just as I said, the solar corona extends as far as the earth.

· THE SOLAR WIND

Along with electrons, there are other particles surrounding the earth, essentially protons, which constitute the solar wind. In the vicinity of our planet, the solar wind blows with a velocity of 450 km/sec and has a density of about 5 protons/cm³. Obviously, if you were to step out of the atmosphere with a hat on, the hat would not fly away; although moving at a velocity of 450 km/sec, 5 protons/cm³ do not have enough energy to blow your hat away. As I mentioned, however, it is the solar wind that blows the gases ejected from a comet away from the sun.

Why is there a solar wind? How come these protons leave the sun and travel in space? I am afraid I cannot give you an exact explanation or any detail about current theories, but I can try to make understandable what happens. If you heat up a gas, it tends to expand.[9] A star is also gas, and very hot. But, as we shall see better later on, a normal star also tends to contract under the pull of its own gravity. Eventually an equilibrium is established between the forces that tend to make it expand and the forces that tend to make it contract. In the corona, however, the velocity of the particles is so high that it corresponds to temperatures of the order of a million to a few million degrees. Under these conditions, the corona tends to expand. In other words, the corona is too hot to remain in equilibrium around the sun. Consequently, the particles constituting the coronal gas continually fly away into space as if there was a process of evaporation going on, and this phenomenon has been given the name of solar wind. If the corona keeps on dissipating into space without pause, it follows that the sun must continually supply it with new material. Roughly, the corona renews itself every day, and this means that the sun loses to space something like 100 billion tons of material every day. Nothing to worry about. If this is the process that will bring the sun to an end as a star, its mass is such that the sun may be expected to last for at least 55,000 billion years.[10]

· THE SOLAR PHOTOSPHERE

First, a few more facts about the sun. If you remember your trigonometry, you can easily calculate the sun's diameter by knowing the sun's distance from the earth and the angle subtended by the sun's diameter.[11] It turns out to be 1.4×10^6 km, or a little more than 100 times the earth's diameter. Hence the sun's volume is about a million times the earth's. From the sun's volume and mass (which can be determined once we know the constant of gravity, the sun-earth distance, and the period of the earth's revolution around the sun), we can calculate that its average density is 1.4 g/cm^3, which is about one and one-half times the density of water and about one-quarter the density of the earth. Observations of the spots that appear on the solar disk show that the sun rotates on its axis, completing a full revolution in about 25 days.

The sun's "diameter," I said. Strictly speaking, one should not speak of a diameter because the sun is a globe of gas with no sharply defined surface separating an outside from an inside. As we proceed inward, however, at a certain point the characteristics of the gaseous mass (density and temperature) change rather abruptly, and the gas, which up to that point was almost transparent to visible radiation (light), becomes opaque; this transition neatly defines a solar "surface." It is to this thin surface layer that we refer when we speak of a solar diameter; we call it the photosphere. This layer is only about 300 km thick, but it is so opaque that all the radiation produced in the sun's interior is completely absorbed before it has a chance to escape. Naturally, no layer, no matter how it is made, can go on absorbing forever. Sooner or later (and I ask my colleagues' forgiveness for such imprecise language) it becomes so full of radiation (that is, energy) that it will either explode or start emitting energy. Under stable conditions, as much energy is absorbed as is emitted. But the energy (radiation) emitted by the photosphere is different from the energy (radiation) it absorbs from the inner layers. Begging my colleagues' pardon once more (better yet, once and for all), let us say that the photosphere behaves like a machine that takes in banknotes of one denomination and gives them out in some other. As much money comes in as goes out, but it is not the same money.

The outgoing radiation emitted by the photosphere corresponds to a gas at a temperature of about 5,700° K.[12] This is why the photosphere is so bright. If we take a photograph of the photosphere with a good telescope and in very good atmospheric conditions (not easy to find anywhere, and certainly not at the bottom of the ocean of air where we usually live), we find that it is not of uniform brightness, but is granular in structure. This phenomenon is called granulation (figure 5). On the average, the individual granules measure 500 to 1,000 km across; they are bright features surrounded by darker, or, rather, less bright, intergranulary spaces. This is what we see when we look at the sun from the outside— hot gas bubbles that rise to the surface with velocities on the order of 300 m/sec, whereas the intergranulary spaces consist of cooled gases subsiding toward deeper layers (figure 6).

This is the classic phenomenon of convection, the same you observe when you look at water heating up in a pot or at the air over sun-heated asphalt. The difference is that the bubbles in the photosphere are as large as France, or the Iberian peninsula, or Yugoslavia, Bulgaria, Rumania, and Hungary put together. Granulation is a never ending process; the individual granules themselves, however, last, on the average, about 5 minutes. In this space of time they form and dissolve, others form and dissolve, and so on, giving the photosphere an ever changing aspect, but always leaving it basically the same. We have known about granulation for a long time, and Janssen had already photographed it in Paris at the end of the last century. Supergranulation, on the other hand, is a recent discovery. The prefix "super" denotes large-scale convective motions in the surface gases; although es-

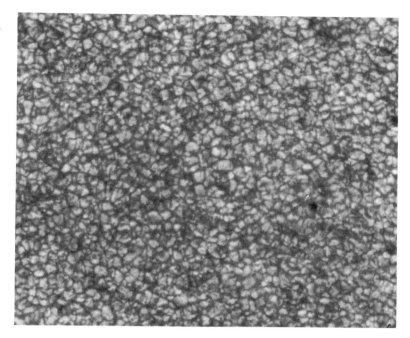

Figure 5
Photograph of solar granulation (June 11, 1967) taken at the Observatory of the Pic du Midi, France, with its 38-cm telescope.

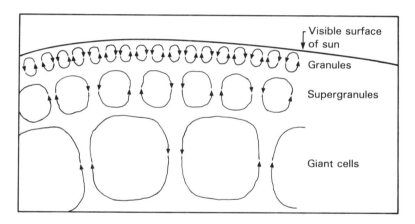

Figure 6
Diagram of convective motions in the photospheric and subphotospheric layers of the sun.

Visible surface of sun

Granules

Supergranules

Giant cells

sentially similar to those we have just described, these motions involve much larger masses of gas that produce longer-lasting granules of remarkable proportions. (They last one day, on the average, and are about 30,000 km across.) As in the case of granules, it appears that the gas moves upward in the central regions of the supergranules and downward at their peripheries. Finally, observations suggest that there are giant convection cells. Each of these phenomena is supposed to correspond to increasingly deeper convective structures: at the top the granules, in the middle the supergranules (at some tens of thousands of kilometers), and in yet deeper layers the giant cells. It is as if the surface convective motions split up into systems of smaller and smaller gas columns or cells as they get closer to the top (figure 6).

· SUNSPOTS

Sunspots also occur in the photosphere, but unlike granules and supergranules they are phenomena associated with the general cycle of solar activity, which involves many other phenomena. Sunspots, in fact, are only the most conspicuous and easily observed signs of solar activity.

Observing the photosphere, we can see that all the spots, or groups of spots (since they generally appear in groups), always cross the sun's disk in an east-west direction and always with the same velocity at a given latitude (figure 7.) Between two possibilities—that the sun stands still and the spots move on it or that the sun rotates and the spots stand still on it (except for small movements within each group)—it is fairly natural to choose the latter. If the spots rotate with the sun, then by measuring the time it takes them to complete a full revolution (and by taking into account that while the sun rotates, the earth moves in its orbit), we can calculate that the sun's period of rotation is about 25 days. It also turns out that the sun does not rotate as a rigid body, but rotates faster in the equatorial regions than at higher latitudes (figure 8).

Sunspots vary considerably in size, but even a small one could easily accommodate the earth (figure 9). Sunspots appear dark because they are cooler regions of the photosphere; the central part of a spot has a temperature 1,000 to 1,500 degrees lower than that of the photosphere. In effect, at temperatures of 4,000 degrees or more, sunspots are fairly bright objects. Thus they

Figure 7
Photographs taken at 1-day intervals of a group of sunspots, showing the rotation of the sun.

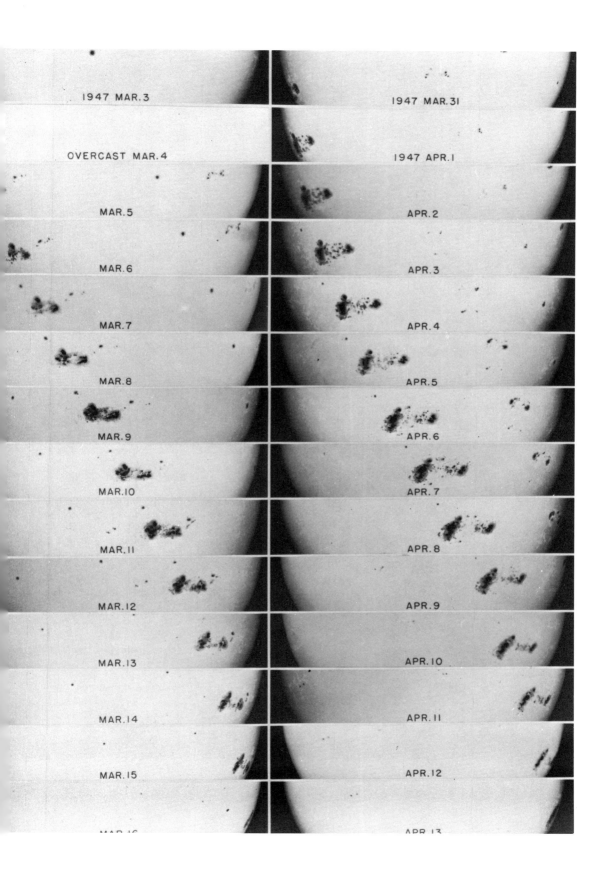

1947 MAR.3

OVERCAST MAR.4

MAR.5

MAR.6

MAR.7

MAR.8

MAR.9

MAR.10

MAR.11

MAR.12

MAR.13

MAR.14

MAR.15

MAR.16

1947 MAR.31

1947 APR.1

APR.2

APR.3

APR.4

APR.5

APR.6

APR.7

APR.8

APR.9

APR.10

APR.11

APR.12

APR.13

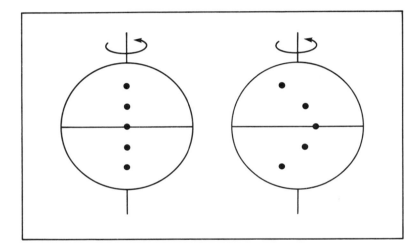

Figure 8
Diagram illustrating that the rotational velocity of the sun varies from latitude to latitude. On the left side of the diagram are drawn five sunspots aligned on the central meridian of the solar disk. After a full solar rotation, the five spots are distributed as shown on the right side of the diagram. From observations of the motions of the sunspots, we may deduce the rotational velocity of the sun at each latitude: at latitude 0° (the solar equator), the period of rotation is 25 days; at latitudes ±10°, 25.2 days; at latitudes ±20°, 25.7 days; at latitudes ±30°, 26.5 days; and at latitudes ±40°, 27.4 days.

Figure 9
The top photograph shows the solar photosphere with a large group of sunspots (April 7, 1947). The earth would appear on this photograph as a tiny disk about 1 mm across, which tells you how truly enormous sunspots are. The bottom photograph shows the same group of spots enlarged many times.

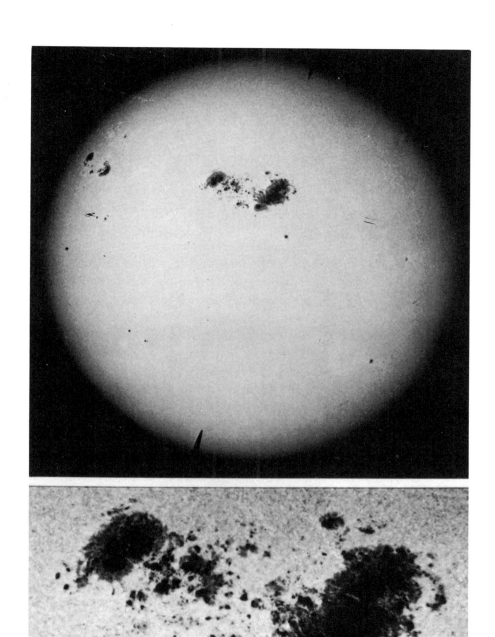

appear dark only by contrast with the brighter areas surrounding them.

Each spot, or group of spots, is rather short-lived, lasting at most a few months, but generally much less.

Finally, sunspots are the seat of strong magnetic fields. You will often read that sunspots are vortexes, but the fact is that despite the many observational data that have been gathered, not much is known about them. One possible explanation of how they form is that the strong magnetic fields associated with them prevent internal gases from flowing outward and producing granulation. When the flux of hot gas is wholly or partly blocked, a region of lower temperature forms. That is a spot. Of the magnetic fields that are supposed to cause sunspots to form, we know close to nothing. Rather, we know many things—direction, intensity, and structure—but nothing certain about their origin (figure 10). Probably there exist intense magnetic fields inside the sun that for some reason and in some way move up to the surface and block all or part of the convection process. This does not happen everywhere, however, since sunspots appear only in a latitudinal band about 80° wide centered on the solar equator; nor does it happen all the time, since for a complex set of reasons, solar activity is a cyclical phenomenon, taking, on the average, about 11 years to go from one maximum to the next.

The local onset of the magnetic field precedes the formation of the spot, and the history of the spot, or group of spots, is only one episode—perhaps the most spectacular—in the history of the magnetic field associated with the perturbation. We do not know how magnetic fields form inside the sun or which mechanisms regulate their movements. In other words, we see certain things happen (sunspots), we are almost sure that they are due to other things that we know exist and that we can measure (magnetic fields), but we do not know why or how the latter form and how they propagate. Sooner or later, perhaps, the skill and imagination of the theoreticians will produce an explanation thoroughly satisfying from both the qualitative and quantitative points of view, but for the time being we must make do with the little we know.

Let us explore the cycle of solar activity a little further. Protracted observations of sunspots show that their number is not constant over time (figure 11). Sometimes the sun appears almost spotless; at other times, it is full of spots. I should point out that we are now considering the phenomenon in its entirety, so that what was already fairly difficult now becomes even more complicated.

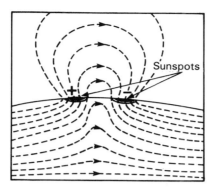

Figure 10
Diagram of the lines of
force of the magnetic field
associated with a bipolar
group of sunspots; + and
− indicate, respectively, the
spot with magnetic polarity
north and the spot with
magnetic polarity south.

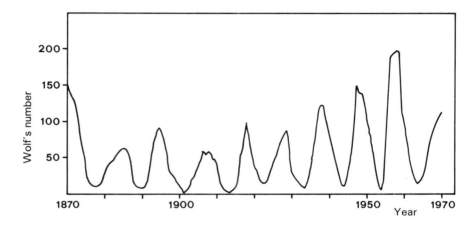

Figure 11
Graph of the 11-year cycles
of solar activity, showing
how the number of sun-
spots varies with time. Plotted on the ordinate is the
so-called Wolf's number,
obtained by multiplying by
10 the number of groups of
spots scattered on the disk
and adding to this the total
number of single spots.

Suppose we start observing the sun during a period of minimum activity (figure 12). There are no spots or maybe a few, once in a while, near the equator. After a while we begin to observe some in both hemispheres, at around latitudes ±40°, hardly any at higher latitudes, and none beyond ±50°. As time goes by, the number of spots increases. Of course, the individual spots disappear, but new ones form in ever increasing numbers, so that the total number becomes greater. The curious thing is that with the passing of time the new spots appear at lower and lower latitudes in both hemispheres, as if there were two rings of perturbation that moved slowly toward the equator. The perturbation, whose existence is revealed by the spots it produces, reaches maximum intensity when the two bands of spots are at around latitudes ±15°; then it begins to subside. As the perturbation as a whole continues to move toward the equator, the number of new spots decreases. When the spots, by now very few, form around latitudes ±5°–±10°, the perturbation subsides. At the same time a new one starts, weakly, roughly at latitudes ±40°, and everything starts all over again. A cycle lasts from 7 to 15 years, 11 on the average. What is it that causes all this to happen? Is there something that slowly rises to the surface and then sinks down again causing the perturbation to become stronger while it moves toward the equator and then subside? Why is it that sunspots do not form at latitudes higher than ±40°–±45° or that the perturbation does not extend to polar latitudes?

Let us complicate the picture even more. A group of spots extends, roughly speaking, along a solar parallel, and its ends have opposite magnetic polarities (roughly speaking again, since both the morphological and the magnetic structures of a group are very complex). During one 11-year cycle all the groups in the sun's northern hemisphere have the same polarity—let us say north—at the western end, while all the groups in the southern hemisphere have the opposite polarity—south in our case—at their western ends (figure 13). In the next cycle all polarities are reversed; the groups of spots in the northern hemisphere have polarity south at their western sides, while those in the southern hemisphere have polarity north at their western sides. In the next cycle there is a new reversal, which brings everything back to the original situation. This means that there is a magnetic cycle of about 22 years.

This so complex panorama is not understood as yet. Not that we lack interpretations; it rarely happens that astronomers lack

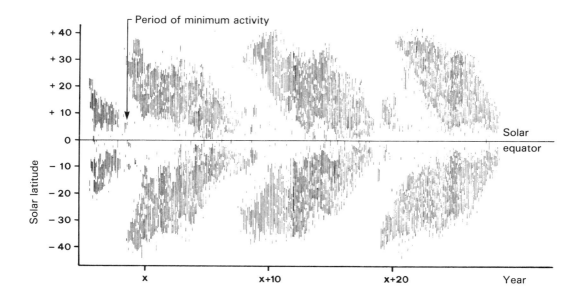

Period of minimum activity

Solar latitude

+ 40
+ 30
+ 20
+ 10
0
− 10
− 20
− 30
− 40

Solar equator

x   x+10   x+20   Year

Figure 12
Maunder's butterfly diagram for three solar cycles, showing the evolution in number and latitude of the sunspots.

Figure 13
The 22-year magnetic cycle of the sun.

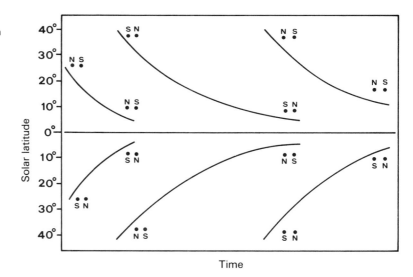

interpretations. While for an observer it is always a joy to look at a star or a nebula, the theoretician's joy is to produce theories. And if a new, unforeseen fact comes up—facts, unfortunately, have a habit of disregarding people's wishes—no problem! The theoretician will put everything to right again: a little magnetic field here, some Alfvèn waves there, a little increase in pressure, a small adjustment in the temperature, another in the turbulence— and everything works out just fine. Not in this case, however. The problem of solar magnetism and related sunspots is a tough one. It is one thing to say that if there is (and there is) a magnetic field in that gas, made in such and such a way, then there can be a sunspot and we understand how it forms; but it is quite a different matter, first, to invent a thoroughly convincing magnetic field that will do all the things we observe and, second, to go on from there to find a mechanism for the production of such a magnetic field.

We shall now see that the problem of sunspots is further complicated by events that occur at different levels and bring us to another phenomenon—the "center of activity"—of which sunspots are only a particular manifestation.

· THE SOLAR CHROMOSPHERE

The state of matter changes considerably above the photosphere; the density is much lower and the temperature rises rapidly. This outer region of the sun, which starts 1,500 km above the photosphere, consists of countless gas jets, called spicules, that rise with velocities on the order of 30 km/sec to heights of 6,000 km and then disappear. On the average, each spicule is 800 km across and only lasts a few minutes. This "burning prairie," as Angelo Secchi (1818–1878) once called it, can be seen by observing the edge of the sun during a total eclipse or with a spectroscope. Figures 14 and 15 show the spicules as they were drawn by Secchi and as they appear in photographs.

By looking at the edge of the sun, we obtain a side view of the chromosphere. But if we now observe the sun in the radiation of much shorter wavelength corresponding to the center of the Hα line of hydrogen[13] (a radiation that we know comes from the chromosphere), we can see the chromospheric structure head-on, as in the case of the photosphere. We can then observe that the chromosphere, like the photosphere, has a seething appearance, with countless structures that move and constantly change shape—

Figure 14
Spicules at the edge of the sun drawn by A. Secchi (left) and as photographed at the Observatory of Sacramento Peak on October 14, 1974 (right).

all of which makes the sun as a whole look less and less like the uniform gas sphere that our grandparents envisioned. Measurements tell us that the dominant factors are the kinetic properties of matter and the magnetic fields that determine its distribution. The chromosphere, too, shows phenomena associated with the solar cycle. I shall mention first one of the more eye-catching features, namely, the faculae, which are bright regions consisting of gas at a higher temperature than the surrounding gas. Actually, the faculae can also be seen projected on the photosphere when they are situated near the edge of the sun's disk, where the photosphere is less bright. Let me explain this point.

The photosphere is less bright near the edge of the sun, or limb, for the simple reason that the sun's temperature increases as the distance from its center decreases. The radiation that reaches the observer from the limb comes from high and relatively cool layers of the photosphere, whereas the radiation from the central region of the disk comes from deeper and hotter layers as well. Hence the limb appears darker. If you look at an image of the sun's disk projected on a screen or at the sun through a very dark filter, you can clearly see that the sun is less bright at the edge than in the center (a phenomenon known as "darkening of the limb"). The faculae are about as bright as the center of the sun's disk, and this is why, when observed directly or, as we say in jargon, in

Figure 15
Region of the sun about
100,000 km across. The
smaller regions defined by
spicules are about 30,000
km across.

white light, they can be seen near the edge but not in the central regions of the disk. But if we observe the sun in the radiation emitted by the center of a strong spectral line, such as the Hα line of hydrogen (figures 16 and 17) or the K line of ionized calcium (figure 18)—radiation that, as I said, comes from the chromosphere (the photospheric radiation is thus excluded)—the faculae can be seen over the whole disk. Although they are not strictly confined to the equatorial region, faculae are very often associated with sunspots and appear in places where groups of spots will subsequently appear at photospheric levels, that is, underneath the faculae, which in general last much longer.

In the areas where sunspots form, quite often, and particularly during periods of maximum activity, we observe small chromospheric regions that in a very short time become remarkably bright. They are known as flares. Like all phenomena of solar activity, they are of short duration—a few hours at most. Seen at the edge, sideways, they look like gas jets rising from the chromosphere toward the corona at fairly high velocities, up to a few hundred kilometers per second. The outbreak of a flare is accompanied by a large increase in the solar radiation in the x-ray, ultraviolet, and radio ranges of the spectrum, as well as by the emission of streams of high-energy particles. It is well-known that this abnormal flux of radiation and particles has various effects on the structure of the earth's atmosphere and magnetic field. It is also well-known that the aurora borealis and aurora australis are due to solar particles emitted by flares that remain trapped in the earth's magnetic field.

·SOLAR PROMINENCES

Prominences are yet another aspect of solar activity. Starting from the chromosphere, they rise to coronal heights. Although generally short-lived phenomena, they may last up to a few weeks or months. In this case they are called quiescent. Active prominences, which are closely associated with sunspots, are of short duration and rapidly changing shape and undergo exchanges of matter with nearby prominences of the same type (figure 19). In a great many cases prominences consist of coronal matter that condenses and sinks to lower levels; in this sense they can be thought of as regions of the lower corona that are cooler than the surrounding medium. Prominences range in size from 50,000 to 150,000 km in height, from 200,000 to 300,000 km in length, and from 20,000

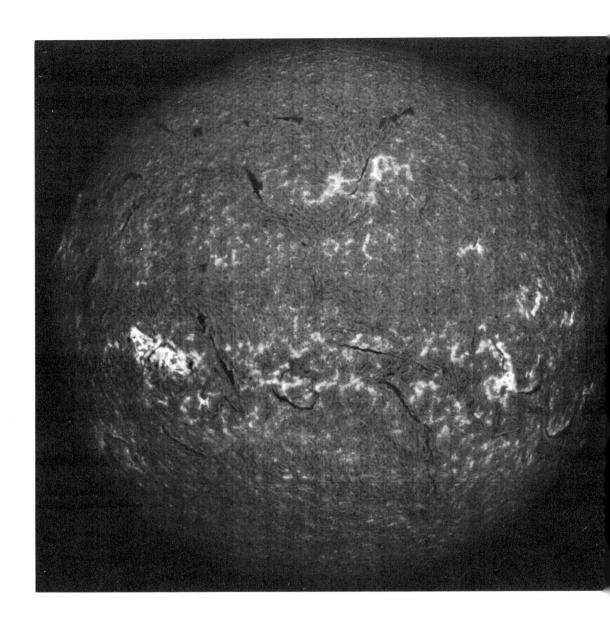

Figure 16
The sun photographed in
the Hα line of hydrogen.
This radiation originates in
the chromosphere. The
photograph (called a spectroheliogram) was taken during
a period of marked activity.
Note the chromospheric faculae (the brightest regions)
and the dark filaments
(which are seen as promi-
nences at the edge of the
sun).

Figure 17
Hα photograph of the
chromospheric region
around a sunspot.

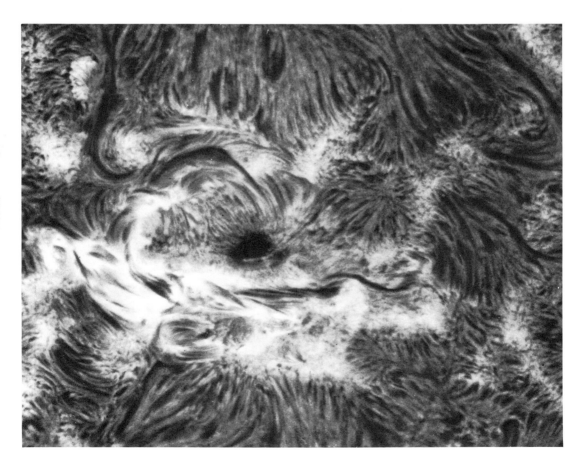

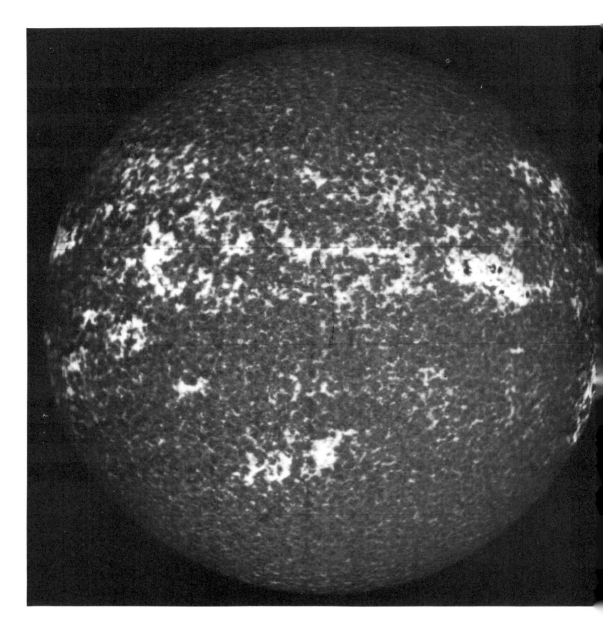

Figure 18
The sun photographed in
the K line of ionized calcium. Notice the faculae
(bright regions). This
spectroheliogram was ob-
tained the same day as the
Hα photograph in figure 16.

Figure 19
Active filament-shaped
prominences near a group
of sunspots.

to 40,000 km in width. The last dimension can be evaluated when prominences, because of the sun's rotation, move from the edge to the face of the disk, where they can be seen head-on. From this perspective they look like filaments, and this is what they are called when seen projected on the sun's disk. At times, for more or less unpredictable reasons, a long-quiescent prominence will erupt (figure 20), ejecting into space an amount of mass of the order of 10 to 1,000 billion tons. It is a lot of mass, but, as usual, the meaning of a lot and a little depends on the unit of measure.[14] Compared with the mass of the sun, which, as you may recall, is 2×10^{27} tons, it is only a billionth of one hundred-millionth of the total mass—nothing to speak of. It is as if a 100-kg man lost a billionth of 1 mg of his weight—something he can do just by scratching his head.

As in the case of flares and all other phenomena of solar activity, the physics of prominences is extremely complex. Let me emphasize that although often done, it is incorrect to describe them as clouds (let alone flames!); clouds calls to mind the formations of water vapor in the earth's atmosphere, while prominences are denser and cooler than the surrounding medium. It follows that they need something to "keep them up." Once again, we have to invoke the sun's magnetic fields, which are quite capable of performing the task.

· TO CONCLUDE

I cannot really blame the reader who has begun to suspect that the sun's magnetic fields have been invented by astronomers to explain a whole assortment of phenomena for which they otherwise could not account. Let me assure you that the sun's magnetic fields do indeed exist and can be measured with techniques ranging from the fairly simple for strong fields to the very refined for weak ones. We can measure their direction, structure, and intensity. Hence, there can be no objection to our using them. As I said, however, much remains to be done in understanding their origin, behavior, and temporal evolution on large and small scales. Furthermore, magnetic fields do not explain every aspect of a group of phenomena as complex as that of solar activity, which involves all of the sun's external layers (photosphere, chromosphere, and corona). Many of these phenomena raise questions that are still far from a satisfactory explanation, and many of the explanations that have been advanced are limited to particular aspects of the

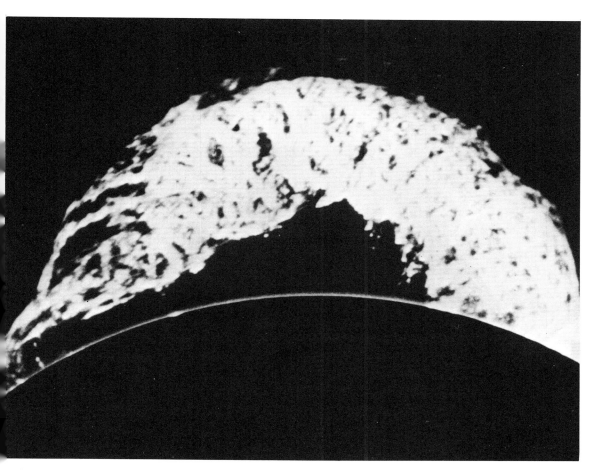

Figure 20
One of the largest promi-
nences ever observed (June
4, 1946). This prominence
rose suddenly from the
edge of the sun and in less
than an hour reached a
height of 500,000 km at a
rate of expansion of
400,000 km/hour.

observed phenomena. The job of the solar physicist is a hard one, for it requires the application of in-depth knowledge of disparate fields to the solution of problems that with time become increasingly more difficult and complex. Is the effort worthwhile? Allow me not to answer this question; I prefer to leave it up to the readers to give their own answers after having read the book.

Let us return to interplanetary space. It is not empty, as it might have seemed. On the contrary, crisscrossed as it is by countless entities—radiation, particles, cosmic rays, dust, micrometeorites, meteorites, comets, asteroids, and planets—it is incredibly crowded. And all these entities are related; they are tied together by mysterious bonds and attractions we call forces or other names; they interact with, influence, meet, hit, and sometimes destroy one another.

• THE PLANETS

The first part of this book is devoted to the earth's neighborhood. Accordingly, I have confined myself to a discussion of the celestial objects in our immediate vicinity. It would not be a bad idea to close this part right now and go on to what happens farther away. I really wish I could forget the planets altogether. First of all, compared with what we find beyond the solar system, they do not count for much. In the second place, there are a great many books that talk about the planets (and tell you, as the saying goes, all you ever wanted to know about them and more), and I am afraid I should be repeating the same old things. But I shall make the effort (and I promise to keep it short) to say something about the earth's traveling companions as well.

I wish that instead of speculating about the kind of life (and especially intelligent life) that might exist on the planets, you would consider the fundamental role that these objects have played in the history of science and humanity. The conquest of the cosmos (intellectual, of course; there has been no other, nor is it likely that there will be any other) started with an understanding of what was happening around the earth. It is obvious that just watching the planets move about in the sky would have never been enough to make sense out of their movements, and in fact it was not. But, inevitably, the day came when their merry-go-round was interpreted in a new way full of momentous implications. This interpretation paved the way for the law of universal gravitation,[15] which is the law that opened up to us the true sky.

And if you wonder what "true" means, I shall tell you that I do not wish to give this word any transcendental significance. True, here, means real, concrete, as opposed to imaginary or invented, as real as the earth we touch with our hands and as real as the moon turned out to be when man finally stepped on its soil.

Thanks to this new interpretation, human beings finally understood their place in the universe. But the most important fact is that this new interpretation produced a new way of thinking—free (at least in principle), critical, blasphemous and iconoclastic if needs be, certainly irreverent. People learned to look about themselves in a different way; they learned to ask questions of nature in the correct manner and to verify the answers. It is thanks to the many observations of planetary motions and to their courageous interpretations that people were able to take to the sky, first to the earth's sky, then to the lunar sky, and now to the skies of Venus, Mars, Jupiter, and Saturn.

Modern astronomy and astrophysics have not added much to the knowledge of planets that classical astronomy had already acquired. When new windows on the universe are opened, our first tendency is to look far away because we can learn a lot more if we study the objects and phenomena whose far-reaching implications provide the elements needed to explain fundamental problems, such as the source of stellar energy and the origin and evolution of stars, galaxies, and the universe as a whole. In addition, strange as it may seem, it is much easier to understand how stars and galaxies are made and work than to make a satisfactory model of a planet. This is because a star, as far as we know and except for a few extraordinary cases, is a gaseous mass at high temperature, and under these conditions the properties of matter are not too complex. (Not that simple, either, and as soon as we go into details the difficulties become enormous.) Consequently, we can make models that, despite their simplicity, are rather gratifying and enable us to explain many facts. When we study the physics of a planet, on the other hand, we deal with the properties of matter at very low temperatures and in condensed states that produce very complicated phenomena. Furthermore, while the life cycle of a star is on the whole fairly simple and similar to the life cycle of other stars, the history of a planet is complex, intricate, and varies from planet to planet. For example, the chemical composition of a planet at a given time is the result of a complex evolution: the influence of cosmic agents on plantary matter, chemical cycles initiated by the sun's electromagnetic and particle radiation, biological processes occurring in surface layers, internal forces, and heating processes that take place in the interior. In the case of the earth, for example, it is almost certain that during its long life there was a period in which, having lost the original atmosphere, it acquired another one because of the release of gas from the interior.

Thus every planet is a different and very difficult problem, and it is no simple matter to reconstruct its history from the data supplied by earthbound observations. It is hard enough to reconstruct the history of the earth, which is under our feet; you can imagine how much harder it is in the case of Pluto, about which we are not even sure of the mass.

There are many things we can do, of course, but it is such a tricky problem that sometimes the result of direct observations are quite different from what had been predicted. To give an example: During a space flight a piece of the vehicle that had taken man to the moon was dropped on the lunar soil, and by registering its impact with special instruments, we discovered that the moon reacted to this very small hit in a totally unexpected manner; it vibrated like a bell for much longer than could be expected from a respectable celestial object. Similarly, we make a lot of conjectures about Mars, and then a space probe shows that we were all wrong. I am not speaking of the canals dreamed up by people who hoped that Mars would be inhabited by beings like us (or possibly better, more clever and intelligent), but in general. You hear that on Mars there is this and that and perhaps that too, but then the probe arrives, takes pictures, sends them back home and, surprise!, it is all different, and there is neither this nor that.

Although we do not know very much about the planets yet, and not much more than was known a few decades ago, we have fairly accurate data concerning such general characteristics as their distances from the sun, masses, periods of revolution, orbital inclinations on the ecliptic plane, periods of rotation, inclinations of their axes on their orbital planes, surface temperatures, and atmospheric chemical compositions. Some of these data are listed in table 1, and I do not need to expand on them because we are all capable of doing so by ourselves. For example, if a planet rotates once on its axis in 48 hours, it follows that a full day there must be twice as long as a day on the earth. And if it revolves around the sun in 2 (earth) years and its orbit is nearly circular—like the earth's—the seasons there must last twice as long as they do here. It is also easy to understand that the apparent size of the sun as seen from a planet is inversely proportional to the sun's distance from that planet. We all know that things look progressively smaller as we move away from them. Funny as it may seem, many authors of popular-astronomy books have such a low opinion of their readers that they explain this fact at great length, filling page after page with useless words when a few lines would suffice. If you place a disk 2 cm across about 2 m from your eye, you will find that it will cover the whole sun (or moon, since both are seen at the same angle). This means that the sun looks like a disk 2 cm in diameter placed 2 m from one's eye. Taking into account the distances of the earth and the other planets from the sun and making a few calculations, we discover some interesting facts (figure 21). From Neptune, for example, the sun looks like a tiny disk about 0.7 mm in diameter placed at a distance of 2 m. Is that the sun? If it were not for the incontrovertible fact that it revolves around the sun, Neptune might have good reason to feel excluded from the solar family. The flux of energy on a square centimeter of Neptune's surface at right angles to the sun's rays is a little over a thousandth of the flux on a square centimeter of the earth's surface. Neptune's days are about as bright as our twilights. It is also rather cold there, considering that the temperature is about 220°C below zero. At this temperature air can become a liquid.

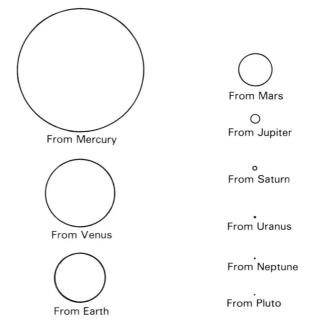

Figure 21
Apparent size of the sun disk as seen from each of the planets. On the earth the sun looks like a disk 2 cm across placed at a distance of 2 m from one's eye (or a 1-cm disk at 1 m, or a 0.5-cm disk at 0.5 m, and so forth). To keep the right proportions, the disks must always be held at the same distance.

From Mercury

From Venus

From Earth

From Mars

From Jupiter

From Saturn

From Uranus

From Neptune

From Pluto

What we know about the planets is enough to convince us that all the bodies, big and small, that revolve around the sun are quite inhospitable, and in this respect space probes have only confirmed what could be easily surmised from past observations. Nowhere in the solar system can we hope to find any sort of life that will make us feel less lonely (if that is the way we feel). Of course, I do not mean lichens or mosses; if such life forms were to be found, say, on Mars (and it is hard to imagine which other planet could harbor anything so remarkable, biologically speaking), it would not surprise a child these days, let alone the scientists who have been expecting it for a long time.

On the other hand, it is entirely possible that in some remote corner of the universe life found the way and the right conditions to start and develop as it did on earth. There is no reason to believe that the earth is a particularly fortunate planet in this respect. The combination of conditions that favor an evolution culminating in biological life—distance from the sun, temperature, atmosphere, and water—cannot be terribly rare. As we shall see later, a star does not always emit energy in the same manner throughout its entire existence; consequently, since life on a planet basically depends on the energy flux that the latter receives from its star, what is possible today on one planet may be possible tomorrow on another. And there is no reason why what is valid for our planetary system should not be valid for all the planetary systems in the universe, and, of course, there is no reason why other planetary systems should not exist.

Before leaving the solar system, I would like to say something about the planets explored by space probes—Mercury, Venus, and Mars—just to prove to my readers that I am quite up to date.

• MERCURY

I have to make a confession. I do not know what prompts me to say—worse, write—such a shameful thing, but here it is (and I am sure many of my colleagues share my guilty secret): I have never seen Mercury (or Uranus, Neptune, and Pluto, for that matter). If I have seen it, I did not know what it was. Furthermore, I have not the least desire to see it, and it is a mystery to me how the ancient astronomers managed to discover its existence. I know that it exists and revolves around the sun; I know that its average distance from the sun is 60 million km, which means that seen from the earth, it is never more than 28° away from the sun and therefore can only be seen—with some difficulty—early in the morning or shortly after sunset; but I feel no compulsion to get up before dawn or to stare at the sky at sunset to see that tiny speck of light and to be able to say, "I have seen Mercury!" Although the ancients did not *know* that this was a planet, they did *see* it. Bear in mind that seeing it once was not enough. They had to see it again and again to realize that that speck of light moved among the stars and therefore belonged to the group of "wondering stars" we know to be planets. Gentlemen, bow low before the ancients!

Mercury orbits the sun in 88 days and, as radar observations have shown, rotates once on its axis in about 59 days, which is about two-thirds of its period of revolution. If we combine the earth's motion with Mercury's, it turns out that Mercury, practically speaking, always shows the same face to us. Since it rotates, however, it does not always show the same side to the sun; hence Mercury, too, has a day and a night. With a mass that ranks it ninth among the planets, Mercury is small and, in addition, is much too close to the sun. As a result, it does not have sufficient atmosphere (if any) to screen out some of the sun's radiation and favor the transfer of heat from the dayside to the nightside, so that temperatures there may reach 350°C (enough to melt lead) on the dayside and −170°C on the nightside. Nasty place.

The fact that it is a nasty place was confirmed by the beautiful pictures taken by *Mariner 10*, which in 1974 passed within 9,500 km of it. At first sight, and if you didn't know any better, you would think they are photographs of the moon (figures 22, 23, and 24). Although Mercury has no satellites, its mass has been calculated from the perturbations it causes in the motions of the other planets and, more recently, from measurements taken by the Venus probes, which passed nearby.

• VENUS

I have seen Venus. Many times. I have also seen it with the telescope, which enabled me to observe its phases as well. The fact that Venus has phases is taken for granted today, but in Galileo's time it was one of the "proofs" advanced to demonstrate that the sun, not the earth, is at the center of the solar system. (And all by itself it was more than enough to invalidate the Ptolemaic system.) Unfortunately, even a momentous discovery tends to lose some of its impact in time, and today we no longer feel

Figure 22
Photomosaic (made up of 18 photographs) of Mercury. The photographs were taken by the television cameras on board *Mariner 10* on March 29, 1974, from a distance of 200,000 km, 6 hours before reaching the point of closest approach. About two-thirds of what is seen in the photograph belongs to the southern hemisphere. The largest craters are about 200 km across.

Figure 23
The southwest quadrant of Mercury photographed by *Mariner 10* on March 29, 1974, from a distance of 198,000 km, 4 hours before reaching the point of closest approach. The largest craters in this view are about 100 km across.

Figure 24
The surface of Mercury
photographed by *Mariner
10* on March 16, 1975,
from a distance of 67,000
km. It is one of the 300
high-resolution photographs
taken by television cameras
on board the probe during
its third approach to
Mercury.

the excitement of the first observations. The night I saw the phases of Venus I slept as soundly as usual.

It does not take much effort to see Venus. Actually, it is hard *not* to see it. Because of its distance and ours from the sun, Venus is never farther away from the sun than 47°. You can see it very easily before sunrise or at sunset and later in the evening as well. At times, depending on the distance and phase, it is so bright that it attracts the eye like an irresistible beacon. Seeing it next to the sickle moon is a lovely sight, particularly when the night is clear and very dark. Since the orbits of the planets and the moon lie in the same plane, such close encounters are not very rare.

Romantic poets, astrologers, and astronomers have paid a lot of attention to Venus, which has been called evening star, morning star, Lucifer, the all-beautiful, the radiant, splendor of the sky, flame, warmth, the luminous, the morning god. But it is only a planet like all the others, and as inhospitable as all the others.

Venus has an atmosphere and is always veiled by an impenetrable cloud cover (plate 3). Until recently nothing was known of its surface, but, as usual, this did not prevent us from speculating. From the amount of solar radiation Venus reflects into space and its distance from the sun, it could be calculated that the temperature there must be lower than on the earth—a hopeful sign in our quest for extraterrestrial life.

Along with the presence of an atmosphere, Venus's size and average density have led us to see it as the earth's twin. Inevitably, with a little contribution from Kant and a lot of help from visionaries, our imaginations populated the planet with manlike beings of superior intelligence and ability. Unfortunately, subsequent measurements of the intensity of radio waves emitted by Venus upset all our calculations, since they implied temperatures of about 300°C and as much as 400°C.

In 1967 *Mariner 5*, a US probe that went around Venus like a motorboat around a buoy, performed experiments on its atmosphere, and *Venera 4*, a Soviet spacecraft, made a soft landing on its soil. *Venera 4* transmitted data for a short time only. Some kind of malfunction, it was thought; everybody knows that these marvelous machines are very delicate. But it turns out that light as it was, it had been crushed by the heavy Venusian atmosphere. The same fate awaited two subsequent probes, *Venera 5* and *Venera 6*, launched in 1969 and built to withstand an external pressure of 25 atmospheres. If they ever touched down, they must have done so as masses of crushed metal. Nevertheless, they made various measurements before reaching levels where the pressure was overwhelming. *Venera 7* was built to withstand a pressure of 180 atmospheres; it landed, radioed back that the pressure was 90 atmospheres and the temperature almost 500°C and after 20 minutes stopped working. It had melted down. In 1972 *Venera 8* worked twice as long. Finally, in 1974 *Mariner 10* passed within 5,800 km of the planet's surface and sent back more than 3,500 photographs of its atmosphere. And now we know enough to conclude that, as far as living is concerned, the earth is a paradise and Venus a hell. Much more interesting from a scientific point of view is the fact that Venus very likely resembles what the earth once was.

Venus's size and mass are about the same as the earth's; hence the force of gravity must be about the same on both planets. At the bottom of its atmosphere the pressure on Venus is 90 atmospheres, and this means that the number of molecules in Venus's atmosphere must be approximately 90 times the number in ours. Since Venus's atmosphere is about as thick as the earth's (measurements tell us that the cloud cover on Venus is 65 km thick), it follows that its density must be greater. Chemically, it consists of about 95% carbon dioxide (versus 0.03% on earth), nitrogen, noble gases, and water vapor. As far as the clouds are concerned, there is some evidence that they may consist of sulfuric acid. There are also traces of hydrofluoric and hydrochloric acids in the gaseous state. The large amount of carbon dioxide and the great density of the atmosphere are directly responsible for the high temperature—the well-known "greenhouse effect." Because of these atmospheric conditions, there must be fairly small variations in temperature between different points on the planet and between day and night. This means that there should not be high winds on Venus, at least at low altitudes.[16] Dense, hot enough to melt lead, and with a pressure at soil level comparable to that that we experience at an ocean depth of 1 km, it is what one would call a "heavy atmosphere"! Hallucinatory too, I might add, because its high density results in a strong scattering of light.

Scattering is the well-known phenomenon that makes our sky look blue. While crossing the earth's atmosphere, sunlight is scattered in all directions by atmospheric particles, but light of shorter wavelength (violet, blue) is scattered far more than light of longer wavelength (yellow, red). As a result, the blue light is deviated from its original path (from the sun to the observer), while the red and yellow light reaches the observer almost undisturbed. What does it all mean? For one thing, since the blue component has been partially scattered, we see the sun slightly yellower than it would appear outside the atmosphere; for another, because of the repeated scattering, we see the blue light coming from everywhere, which is what makes the sky look blue. If there were no atmosphere, the sky would look black, and, in fact, the higher we go the darker it looks, as you can see from a plane or a mountain top. If the density of the atmosphere increases or the light has to travel much longer through it, the effect becomes increasingly more conspicuous, and at a certain point even light whose scattering was once negligible becomes greatly scattered. It follows that when the density is high enough, sunlight will no longer reach the observer directly but from all parts of the sky (an effect we observe here on earth when we are enveloped by fog or smoke). We see light, but not the sun, and this happens whether there are clouds or not. The same happens when we look through the atmosphere from the outside. The scattering of sunlight prevents us from seeing anything that is located below a certain depth. This is the reason why Venus's soil is invisible to outside observers and, conversely, why the Venusian sky cannot be seen from Venus's soil, clouds or no clouds.

On Venus, the sun should be barely visible as a faint reddish halo. The absorbent layer of clouds must make visibility at soil level even worse; and even when the sun is high on the horizon, there cannot be more light there than we have at dawn. The strong scattering has an additional effect, somewhat similar to the effect created by a fog. The scattering of violet light is such that a violet source becomes invisible at a distance of a few dozen meters. A red source should be visible up to a few hundred meters. Consequently, to have any sort of horizon in an atmosphere of this type, one would have to look through red filters.

Is there any hope of finding out what the Venusian soil looks like? Actually, there is a way. As I said, scattering becomes less appreciable as the wavelength of the radiation becomes longer. Radio waves are just the right length, and efforts are currently under way to explore Venus with radar techniques. We have already learned a few things, such as the existence of mountains (figure 25), and in a few years radar surveys should give us maps of Venus as detailed as any lunar map obtained by optical telescope.

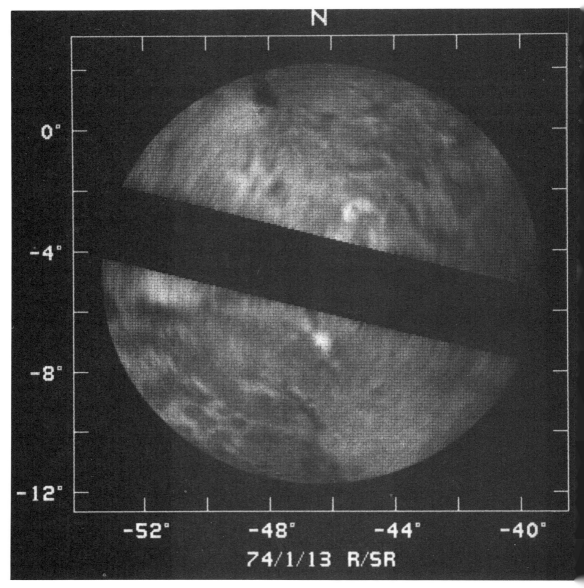

Figure 25
Radar map of a region of
Venus 1,500 km across
near the equator. A white
spot adjacent to a dark spot
indicates an elevation; the
bright part is one slope and
the dark part the other. The
photograph was taken with
a spatial resolution of
10 km by 10 km (which
means that details smaller
than this cannot be seen).
The dark band is a region of
low resolution that has been
erased.

·MARS

Let us move on to Mars (figure 26). After all that has happened in recent years everybody knows everything there is to know about Mars, and it seems useless to write about it. I shall confine myself to a few comments that might prove useful.

People have been interested in Mars since at least 1877, that is, since Schiaparelli thought he had discovered a network of canals on it so regular in appearance as to be the work of intelligent beings. Telescopic observations of seasonal variations, such as the disappearance of white polar caps—surely ice—and the concurrent appearance at lower latitudes of green regions—surely vegetation—were interpreted as evidence of the existence of an advanced race in its death throes. Mars, so the story went, was a dying world on which water had become very scarce; in winter water somehow collected at its poles, forming the ice caps seen with the telescope; in spring the water melted and was carried away by canals, sometimes thousands of kilometers long, to every corner of the thirsty world, which would promptly become green again. Pale springs and cheerless summers of an intelligent race fighting for survival! Of course, to be able to build such an impressive irrigation system, those creatures had to have reached a level of civilization that we poor sods could not even dream of.

Since then, in different ways and with varying degrees of foolishness or reasonableness, Mars has been the object of much study and speculation, which was not lessened in the least by the discovery that the canals were mere optical illusions.

The year 1877 also saw the discovery of the two moons of Mars. It is funny to think that Kepler should have predicted their existence on the basis of arguments that today would not convince a child; and funny that in 1726 Jonathan Swift should write in *Gulliver's Travels* that Laputa's scientists had discovered that Mars had two moons—Phobos and Deimos—orbiting the planet in 10 and 20.5 hours, respectively. The actual periods—7^h 39^m and 30^h 18^m (for notation, see the note to table 1)—are not too different from those invented by Swift.

Mars's moons are quite small and irregularly shaped—just two boulders, some 20 and 10 km across, that can barely be seen even with powerful telescopes. In 1944 the combined results of past observations seemed to suggest that Phobos's orbit was getting smaller and that the satellite was speeding up. After various hypotheses were discarded because of inconsistencies, the possibility

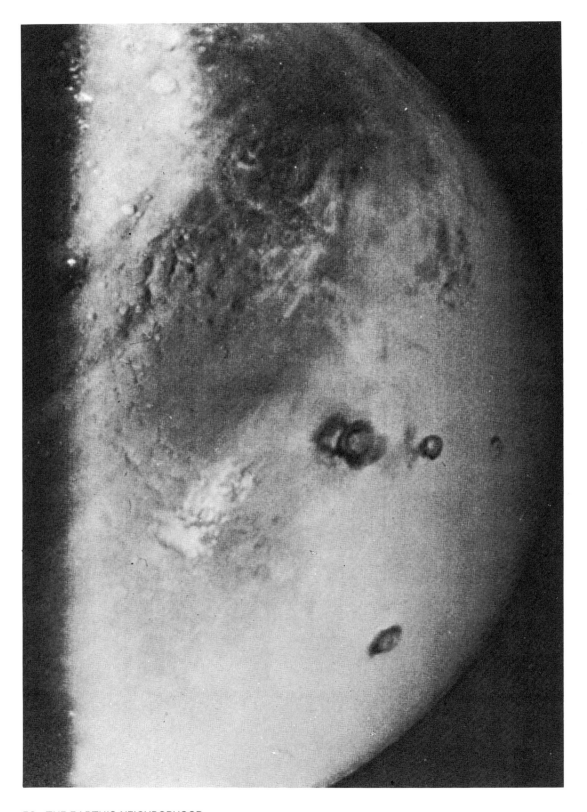

was considered that Phobos's motion might be affected by Mars's atmosphere. The trouble was that Mars's atmosphere is too thin to exercise sufficient drag. This is true, however, only if Phobos if rather dense, like all the other planetary objects, and hence of considerable mass. But what if its mass is small? In this case, since the volume is what it is and cannot be changed, the density has to be very low. Atmospheric drag could conceivably affect the motion of an object of small mass. By calculating backward, it was found that to justify the breaking action of the Martian atmosphere, Phobos has to be a thousand times less dense than water. Since it is practically impossible to conceive of a material of such a low density, the hypothesis had to be abandoned. If at the time of reckoning a theory does not square with the facts, there is no choice but to throw it out. Too bad, just another false start. It is all part and parcel of the researcher's job and not such a disaster at that. It is not entirely without value to prove, say, that a road presumably leading to Rome does not actually get there. Certainly any Rome-bound traveler cannot but find it an interesting piece of news; it does not quite solve the problem of getting to Rome, but it eliminates one of the possible errors.

What then? Back to the Martians, of course. If you imagine an object 20 km across to be hollow, then the mass becomes small and the average density (mass divided by volume) becomes very small. Since things like that do not exist in nature, it means that "somebody" must have made Phobos. In an age that has seen the launch of the earth's man-made satellites, the idea of a giant Martian-made satellite was bound to come up. The Martians might not have built canals, but now it seems that they built satellites that are far beyond our capabilities. On the other hand, Mars shows no trace of all the mighty works that should be there. Hence, so this logic goes, the Martians are gone; they lived, they were great, they reached the highest level of technological development, but in the end they were defeated by a hostile nature. They are now extinct, and it is a great pity that we should have found brothers in the cosmos too late. It is such a great pity that many people refuse to accept the evidence and continue to believe in their own dreams. Some even see Martians on the earth. There was a time, a few years back, when flying saucers were sighted everywhere. Eyewitnesses reported not only to have seen flying saucers land but to have been invited or dragged inside by the creatures operating them. Today there do not seem to be so many saucers around; undaunted, the believers keep on looking, form associations, and publish silly papers replete with mysterious happenings. Since these papers sell, somebody must be buying them. But there is nothing remarkable in that. In our consumeristic society anything sells. There are so many of us that there will always be found people ready to agree enthusiastically with anything we say.

Coming back to Mars's moons, various other hypotheses were advanced, including the possibility that the apparent change in Phobos's motion might be the result of faulty measurements. It was also suggested that the two satellites might be asteroids captured by Mars or (according to a theory about Mars's genesis that is not worth discussing) fragments of Mars itself. Then again, it might be this or it might be that. This is usually the way it goes in astronomy. We theorize, we speculate, we argue, we conjecture, we talk, we think. Sometimes, unfortunately, there is no other way and it cannot be helped. But once in a while we have the chance to see with our own eyes, and then we almost always find that one observation is worth a thousand arguments. This in no way means that speculation is useless, but simply that experience is fundamental and that Saint Thomas's attitude was basically sound. The best proof of this is in the work of Galileo Galilei, whose simple observations shattered all the crystal spheres the ancients had built in the sky. In our case it was the Americans who provided the evidence by launching the space probes of the Mariner series, equipped with cameras. In 1965 *Mariner 4* photographed some regions of Mars that turned out to look like moonscapes. In 1969 *Mariner 6* and *Mariner 7* photographed different regions, as well as the planet as a whole, and enriched our view of Mars by recording plains, valleys, and flats. In 1971 *Mariner 9* discovered volcanoes and riverbeds and performed topographical surveys of the whole planet that revealed mountain ranges, gullies, and large canyons. One Martian canyon is 7,000 m deep and long enough to cross the United States from the East Coast to the West Coast. One extinct volcano rises 26 km above the surrounding plane. Another one, Nix Olympica (figure 27), is about the same height and measures 500 km across at the base (equivalent to the distance between Naples and Florence); at the summit, its crater is 70 km wide. *Mariner 9* also photographed Phobos and Deimos, which—I daresay "naturally"—turned out to be what good sense dictated, namely, two big rocks pitted with craters of all sizes due, like craters on the moon, the earth, Mercury, and Mars, to the impact of meteorites and similar objects that in a distant past

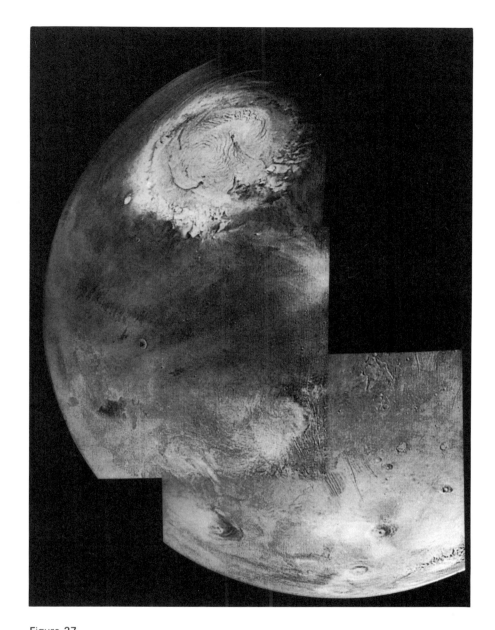

Figure 27
The northern hemisphere of
Mars photographed on Au-
gust 7, 1972, by *Mariner 9*
from a distance of 13,700
km. Observe the polar cap
(top), the volcano Nix Olympica (bottom left), and the
last piece of the great equatorial canyon along with
other volcanoes (bottom
right).

struck the inner part of the solar system like a spray of bullets. So much for the Martians.

And then came July 1976, or more precisely, July 20, 1976 (another date we cannot forget); on that day *Viking 1* landed on all fours upon the Martian soil (plate 4). The historic event was duly covered by the mass media, and in the words of some enthusiastic commentators the whole world waited with bated breath to learn whether there was life on Mars.

In truth, it was a fantastic achievement. Aside from the cost (more than $1 billion for both *Viking 1* and *Viking 2*, which was to land in September), it took 15 years of work, research, and technological development to achieve the soft landing of a spacecraft loaded with delicate scientific instruments on a target 380 million km away after a 10-month, 800-million-km journey through space—a craft, moreover, that had to maneuver automatically in the final landing stages, since it could not be guided from the earth by radio signals, which would have taken too long to come (about 20 minutes from the earth to Mars).

Mars is not a dead planet. It is geologically active, although less so than the earth (but more so than the moon). Its atmosphere is very thin (though the winds there can be fierce because of the large variations in temperature) and consists essentially of carbon dioxide plus some carbon monoxide, water vapor and ice, and traces of ozone and possibly other gases. In summer the surface temperature at the Martian equator is around 28°C shortly after noon and more than 50°C below zero before sunrise. At the poles it is −90°C.

Needless to say, it is not an environment suitable for human beings, and we must resign ourselves to the fact that we are alone in the solar system. It is sad. But look at it this way. There are more than enough of us here on the earth, and we are always fighting with each other as it is; we may just have missed the chance to fight on a planetary scale. Actually, there were not too many people left who still believed that Mars might harbor some complex form of life. Observations made from the earth with telescopes equipped with special instruments, such as spectroscopes,[17] had already told us that Mars's atmosphere was mostly carbon dioxide, and we already knew that Mars's temperature was low and subject to very large variations. All we could hope to find was some evidence of life, past or present. We would have been happy to find traces of the kind of life that here on earth, where human life itself is not very valuable, we never even notice.

So far, however, despite the fact that the Viking probes have done a good job of analyzing the soil samples taken by their mechanical arms, no trace has been found of those organic compounds that are the building blocks of life. But we have not given up hope yet, and if these probes cannot settle the matter, perhaps others will try.

•BEFORE WE MOVE ON TO SOMETHING ELSE

I have told you something about the planets that have been explored by space probes because these are events that unfold under our own eyes. It may also be worth discussing space exploration itself, or to put it less arrogantly, the exploration of nearby celestial objects, which is the only celestial exploration we can accomplish. The surprising thing about all these enterprises is the close involvement of the military, which, truth be told, has never shown a great love for pure science. Better late than never, you will say; much better that the military should spend its money on scientific research than on making war. To be sure. But there are many who fear that the very advanced technology we have developed to go to Mars and to analyze its soil might be used, with a few appropriate modifications, to make terrifying weapons. And there are many who believe that the owners of these marvelous toys might intimidate the whole world into doing anything they pleased. On the other hand, there is nothing we can do about it. The world is propelled by social, economic, and political forces whose motivations in most cases are only evident after the deeds are done.

With regard to space science, it is natural, or at least understandable, that most astronomers should not worry too much about the consequences their work might have outside their own fields. They might be more sensitive to, and critical of, any biological research that represented a potential threat to human beings. But in this case it would be the biologists who would try to play down the dangers, since it is almost inconceivable that there are scientists willing to give up their research, particularly when it shows promise of fundamental breakthroughs.

For this reason, the old adage that war is too important to be left to the generals should be applied to science, which is too important to be left to the scientists. Actually, science has not been left to the scientists for a long time now because it is the politicians and the military who have been financing their research and setting their priorities. The military does not make the best partners or advisors—or, considering actual practice, bosses—for scientists. I think that every citizen ought to take an interest in scientific research, so that in the future decisions may be influenced by "public opinion." For this to happen our schools must awaken a scientific awareness and all professionals must fight the kind of misinformation and charlatanism fostered by books, articles, and television programs whose only aim seems to be to show science as a mystical experience better left to a brotherhood of initiates.

To come back to space exploration, I think that it could and should have been done with far less urgency and as part of a long-range plan for the advancement of human knowledge—a plan that should have taken into account humanity's most pressing problems in their order of priority.[18] And the existence of life on Mars is certainly not one of them.

I trust it is clear that I do not consider space flights the scandal of our time. There are many other things more deserving of the name. But it is also true that space flights share the same faulty logic. They are a symptom, one of many, of the sickness of our world.

Let us hope that we shall recover.

PART II
THE GALAXY

Sky, and naught else: the deep dark sky
Strewn with great stars; the sky, where
All that was earthly appeared submerged.

And the Earth I perceived in the Universe.
Quivering, I felt that She too is of the sky.
And I saw myself down here small and lost

Wandering on a star, among the stars.

Giovanni Pascoli
From ''The Fireball''

•BEYOND THE SOLAR SYSTEM

The solar system is not infinitely extended. The ancients had come to the same conclusion, but they could do no better than to delimit it with a "sphere of fixed stars" beyond which was the Void. Without really wanting to or being aware of it, Copernicus started a revolution that had been brewing for a while and that through the work of Tycho Brahe, Johannes Kepler, and Galileo Galilei, to mention only the most famous names, was to lead to Newton's law of universal gravitation. The finite world of the ancients was shattered, to be replaced by an immensely larger universe. Of course, it did not happen in the blink of an eye. Copernicus was born in 1473, and the first edition of his celebrated *De revolutionibus orbium coelestium* appeared in 1543. Galileo published his *Sidereus nuncius* in 1610 and in 1633 was condemned by the Church. Giordano Bruno was burned at the stake in Campo dei Fiori in 1600. Newton wrote his *Principia* in 1687. I shall not go into a detailed historical analysis of the Copernican revolution, interesting as it might be, because it goes beyond the scope of this book, which is much more modest: to discuss a number of celestial phenomena in a manner as far from pedantic as possible and to give a feeling for the world we live in rather than an exact account of everything.

Let us look at the sky without preconceived notions. If there is no sphere of fixed stars, how is the sky made? What are those specks of light we call stars? Are they near or far? How many are they? Are they scattered at random and more or less uniformly distributed through space or do they follow some sort of pattern? And if so, what *is* the pattern? Also, now that space is no longer defined by crystal spheres, we would like to know how large it is and whether it is a perfect, never changing, absolute, infinite entity, or does it, too, have unforeseen properties we can discover and reason about.

•MEASURING STELLAR DISTANCES

Looking at the sky without preconceived notions means that we are willing to accept results that are totally unexpected and even contrary to common sense if they are substantiated by experience and reasoning. An open mind is one of the first benefits we can bestow on ourselves.[19] The second is the entire arsenal of instruments, mathematical techniques, and theoretical models that science and technology have been able to devise and to develop to date.

One of our first questions is, How far away are those specks of light? That the moon, sun, planets, and comets are relatively nearby is almost obvious. Rather, it is entirely obvious. If we observe the sky for some length of time, we soon notice that they move with respect to the stars, which instead appear fixed in the heavens, each in its own place. The constellations, whose patterns are drawn by stars, do not change—proof that the stars do not move in relation to one another—nor would they change if we were to watch them for all the years of our lives. It follows that the objects of the solar system are near and that the stars are far away—very, very far away. Recall what happens when you travel in a car or train; nearby objects appear to move by very fast, while distant things move gradually more slowly. The stars appear stationary; hence they must be frighteningly far away.

There was a time when this line of reasoning did in fact frighten people because of its implications. When Copernicus and his followers proposed the planetary model we now call the Copernican system (with the earth orbiting the sun), one of the objections to it, raised by Tycho Brahe himself, was the question of the apparent motions of the stars, which is a necessary consequence of the model. If we travel with the earth around the sun, then our line of sight to any given star must change in the course of the year in such a way that the star must appear to move in a small circle. Attempts were made to observe this effect but were unsuccessful because the instruments of the time were not good enough to make precise measurements. This left two possibilities: either the earth did not move around the sun, or the stars were very far away—one could say infinitely far, since the diameter of the earth's orbit was a point in relation to stellar distances. At that time people were not prepared to believe that the universe was as vast as needed to account for the lack of apparent motions, while they had very good reasons to believe that the earth stood still. As a result, they went on thinking in the old way.

Aside from apparent motions, which are only an optical effect, the stars do not seem to move. This raises the same kind of question that troubled Copernicus's contemporaries. Are the stars really stationary, or are they so far away that we cannot see them move even over long periods of time? And just how far away are they?

None of these questions can be answered so long as astronomers have to rely on the naked eye and fairly primitive instruments. This is because the quantities that have to be measured (angles) are very small and require for their accurate measurements more sophisticated devices. We shall deal shortly with the proper motions of stars. Right now, let us see how the distance of a celestial object is determined.

The method is simple, and if you know plane trigonometry you can figure it out by yourself. It is the same method the surveyor uses to solve one of his classic problems: Determine the distance of an inaccessible object (on the other side of a river, for example) by means of an instrument that measures angles and one that measures lengths. The former is a theodolite; the latter, a measuring rod. Figure 28 illustrates the surveyor's problem. The surveyor is at point A and measures the angle α subtended by the lines AO and AB, where B is a point arbitrarily chosen. Then moving to point B, he measures the anble β subtended by the lines BO and BA. Last, he measures the distance between B and A, that is, the segment AB. This done, he can sit at his desk and apply the rules of trigonometry, which allow us to calculate any element (side or angle) of a plane triangle when three of its six elements are known. In particular, he can measure the angle at O (it is 180° minus the sum of α and β) and in short order from this the distance between A and O. If you doubt the result, the surveyor can easily prove to you that the method works by performing the same operation for a distance AO that can later be verified.

In the case of stars, the baseline AB is given by two points of the earth's orbit and the rest follows almost immediately. I said "almost" because things are not quite as simple for astronomers

Figure 28
Determination of the distance AO of an inaccessible object O.

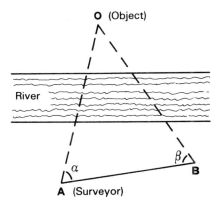

as for the surveyor, in that we cannot aim our theodolite (which, for astronomers, is a telescope) first at a star and then at another point of the earth's orbit. As a result, our method is more indirect, although basically the same. Forgive me if I do not explain it. This is not a book on observational techniques, after all, and a quick explanation might be worse than none. The important thing is that the principle should be clear; if you are really interested, you can always find out what the actual procedure is.

The first star whose distance could be determined was 61 Cygni. The measurement was made by F. W. Bessel in 1838, and the distance turned out to be 10.5×10^{13} km, or a little more than 10^{14} km. I shall not insert here an exclamation point, but you should think about it a moment; 10^{14} is a very large number. Now you can understand why we had to wait so long to obtain it and why even Tycho Brahe, the best observer of his time (indeed, the best observer up to his time), could not do as much. As I said, we have to measure angles. As the distance of the object O in figure 29 becomes larger and larger, the lines AO and AB tend to become parallel, α and β each tends to 90°, and the angle at O we need to measure in order to calculate the distance becomes smaller and smaller. In the case of 61 Cygni this angle is 0.586″,[20] or a little more than half a second of arc. It is the same angle that would be subtended by a ball half a meter across at a distance of more than 700 km. The angle 0.586″ is the largest that can be measured for 61 Cygni and is obtained by taking the diameter of the earth's orbit as baseline. Half this angle, 0.293″, is called the parallax of 61 Cygni. In general, the parallax of a star is the angle from the star when the baseline is the radius of the earth's orbit. When this angle, or parallax, is 1″, the star is at a distance we call a parsec. One parsec corresponds to about 3×10^{13} km, or 3.26 light-years, a light-year being the distance light travels in one year at the well-known speed of about 300,000 km/sec. The reciprocal of the parallax, which is in seconds of arc, gives distance in parsecs. A parallax of 1″ corresponds to a distance of 1 parsec, or 3×10^{13} km; a parallax of 0.1″ corresponds to 10 parsecs, or 3×10^{14} km (a distance 10 times greater); a parallax of 0.01″ corresponds to 100 parsecs, or 3×10^{15} km; and so forth. The nearest star, Proxima Centauri, is at a distance of 1.33 parsecs (4.34 light-years) and has a parallax of 0.76″, which is the largest parallax measured for any star.

It is now evident why astronomers in centuries past could not measure stellar distances by the trigonometric method. I should

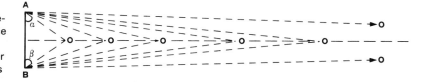

also point out that measurements of the type Bessel made furnished
a direct, incontrovertible proof that the earth revolves around the
sun.

The reason why I have lingered on this matter is that these
direct measurements are the foundation stone for all the rest, that
is, for all the indirect methods that must be used when the direct
approach is not possible. The indirect methods must also work
for stars to which we can apply the direct methods; or, as we say,
the indirect methods must be calibrated to the direct ones.

Once it is understood that the universe is infinitely larger than
the ancients imagined, it is not surprising to find that there are
only a few stars whose parallaxes can be measured by the trigonometric method.[21] In addition, when we perform such measurements, of necessity we make *errors*. I am aware that this word
could be the start of a philosophical discussion, since saying that
we make errors in measurement seems to imply that we believe
that there exist intrinsically *true* measurements, but I refuse to be
drawn into it. Discussions of this type are not serious anyway
and should be left to philosophers who have not studied physics
(a common occurrence, unfortunately). Let us just say that for a
number of reasons different measurements, made, no matter how
carefully, at the limit of an instrument's capability necessarily give
different results. It is in the nature of things. To do things properly,
therefore, every measurement must be followed by its so-called
error,[22] which is obtained by taking several measurements and
evaluating the results appropriately. In the measurement of angles,
the error we make today with the most sophisticated techniques
and the best instruments available is ±0.02″. This means that by
the trigonometric method we cannot make exact measurements
of parallaxes smaller than a few hundredths of a second of arc,
equivalent to distances greater than 50 to 100 parsecs.

To conclude, the stars are extremely far away from the sun—
and hence from each other, since it is inconceivable that the sun,
and only the sun, should be isolated in space. In scale, if the sun
were a sphere 10 cm across, the nearest star would be 2,800 km
away. Think about that.

Let us return for a moment to the solar system. The distances here are not nearly so awesome and can be measured by taking as baseline, for example, two points on the earth. There is another way of measuring distances within the solar system that does not concern us right now. But a simplified version of it allows us to calculate all these distances when we know one of them. It is based on Kepler's third law, which states that the distances of the planets from the sun are related to their periods of revolution (easily obtainable by observation).[23] We can thus make a scale model of the entire solar system, and knowing one distance, we need only multiply all the others in the model by the scale factor to obtain the actual distances of the individual objects from the sun. To obtain very accurate measurements we use the nearest possible objects, which are not the planets, but certain asteroids whose very eccentric orbits bring them closer than the planets to the earth. An asteroid (one of those normally located between Mars and Jupiter) that became famous in this respect is Eros, which in 1900–1901 passed within about 40 million km of the earth and in 1930–1931 as close as about 25 million km—much closer than Venus and Mars, as you can see from table 1. In this manner we can determine the earth's distance from the sun,[24] which is approximately 150 million km, or about 100 times the sun's diameter.

Today there is another, more accurate, way of measuring distances based on the measurement of times instead of angles. It involves sending a radar signal to some target and measuring the time elapsed between its departure and the arrival of its echo. The propagation speed of electromagnetic waves (the speed of light) is known with great precision, and by multiplying this speed by half the time it took the signal to go and to come back, we obtain the distance of the target. Very simple.

·CAN WE LEARN MORE?

Back to the stars. We have learned to measure the distances of the nearest stars, and this is certainly a fundamental step. But now we would like to know whether it is possible to determine the distances of stars that are more than 50 to 100 parsecs away (and hence beyond the capability of our instruments) and whether we can learn other things about them as well, such as their masses, temperatures, sizes, structures, the amounts of energy they emit, the transformations they undergo (if any), and their motions in space. The answer is, Yes, we can. Since it would not be much fun for you to read or for me to write a series of affirmatives, I shall expand on them as simply as I can, and I hope you will forgive the inevitable imprecisions and simplifications. I believe that it is more important to explain the concepts than to go into minute detail. My aim is to make understandable in a general way how astronomers work, how they make their statements as correct as possible by separating what is certain from what is uncertain and then by evaluating the degree of uncertainty. I also wish to show that astronomical research does not need to be surrounded by a magical halo. Like all sciences, astronomy is a human activity, done by human beings for human beings,[25] and it is wrong to present it in such a way that the reader, overawed by the sublime activities of modern sorcerers, will humbly withdraw from the scene—very impressed, but as ignorant as before.

•STELLAR MAGNITUDES

Let us begin with luminosity, the amount of energy a star emits every second. Before we discuss luminosity, however, we must introduce other concepts, and first of all the apparent magnitude.

If you look at the sky on a clear night you see lots of stars—between 1,000 and 3,000 (depending on your eyesight, the lights around you, and whether the moon is up). Apart from the number, anyone can see that some stars are brighter than others. Thus we can grade them in order of brightness and establish what is called a magnitude scale. The brightest stars shall be assigned to the first class; stars that are half as bright shall make up the second class; and proceeding in this way, we can establish additional classes. It is a fairly rough classification, since our eyes are not nearly accurate enough, but it is better than nothing. The ancients, who made this classification, on the whole did pretty well with it. How many classes can be established? The answer is six. The sixth class is made up of stars that can only be seen by the most acute eyes; and if you are a little shortsighted, you will not be able to see fifth-magnitude stars either. Never mind; the sky is just as beautiful without them. Modern measurements (not done by eye) show that sixth-magnitude stars are 100 times less bright than first-magnitude stars. This means that by the unaided eye the ancients assigned to a class those stars that, on the average, are 2.5 times less bright than stars of the preceding class. Note that the fainter the star, the larger the value of magnitude. (This

TABLE 1
Some data on the sun and its planets

	Sun	Mercury	Venus	Earth
Equatorial diameter (km)	1.4×10^6	4,880	12,104	12,756
Mass[a]	3.35×10^5	0.055	0.815	1
Volume[a]	1.3×10^6	0.06	0.88	1
Mean density[b]	1.4	5.4	5.2	5.5
Surface gravity[a]	28	0.37	0.88	1
Maximum magnitude	-26.8	-0.2	-4.22	—
Mean temperature (°C)	5,800	350; -170[c]	480; -33[c]	22
Atmosphere (principal constituents)	H He	—	CO_2 N_2 O_2	N_2 O_2
Period of rotation[a]	27	59	243 (retrograde)	1
Mean distance from the sun[a]		0.387	0.723	1
Mean distance from the sun (10^6 km)	—	57.9	108.2	149.6
Period of revolution[a]	—	0.24	0.62	1
Period of revolution	—	88 days	224.7 days	365.26 days
Mean orbital velocity (km/sec)	—	47.9	35.0	29.8
Inclination of the equator on the orbit	—	$<28°$	3°	27°27'[e]
Inclination of the orbit on the ecliptic	—	7°00'	3°23'	—
Eccentricity of the orbit	—	0.206	0.007	0.017
Number of satellites	9	—	—	1

a. Each figure is given relative to the earth; the absolute value is obtained by multiplying the relative value by the data given for the earth, which are listed in a separate table.

b. Density of water = 1.

c. The two temperatures refer to, respectively, the light and the dark sides of the planet.

d. The constituents abbreviated are hydrogen (H), helium (He), carbon dioxide (CO_2), nitrogen (N_2), oxygen (O_2), ammonia (NH_3), and methane (CH_4).

e. $1' = 1$ minute of $1° = 1°/60$; $1'' = 1$ second of $1° = 1°/(60 \times 60) = 1°/3,600$.

Mars	Jupiter	Saturn	Uranus	Neptune	Pluto
6,787	142,800	120,000	51,800	49,500	6,000?
0.1075	317.83	95.147	14.54	17.23	0.17
0.15	1316	755	67	57	0.1?
3.9	1.3	0.7	1.2	1.7	?
0.38	2.64	1.15	1.17	1.18	?
-2.02	-2.6	$+0.7$	$+5.5$	$+7.85$	$+14.9$
-23	-150	-180	-210	-220	-230?
CO_2	H_2 He CH_4 NH_3	H_2 He CH_4 NH_3	H_2 He CH_4	H_2 He CH_4	?
1.03	0.41	0.43	0.45 (retrograde)	0.65	?
1.523	5.202	9.539	19.182	30.058	39.44
227.8	778.3	1427	2869.6	4496.6	5900
1.88	11.86	29.46	84.01	164.79	247.7
687 days					
24.1	13.1	9.6	6.8	5.4	4.7
25°59′	3°05′	26°44′	82°05′	28°48′	?
1°51′	1°18′	2°29′	0°46′	1°46′	17°10′
0.093	0.048	0.056	0.047	0.009	0.250
2	13	10	5	2	0

Data for the earth	
Mean radius (km)	6,371
Mass (g)	5.976×10^{27}
Volume (cm³)	1.08×10^{27}
Surface gravity (cm/sec²)	9.81
Period of revolution (with respect to the fixed stars)	365.256 sidereal days
Period of rotation (with respect to the fixed stars)[a]	23^h 56^m 04^s
Mean distance from the sun (km)	149.6×10^6

a. 23^h 56^m 04^s is shorthand for 23 hours, 56 minutes, and 4 seconds.

holds even for movie stars, among whom "a star of the first magnitude" is one of the most famous, while "a star of the second magnitude" is one of the less celebrated.)

Unlike the ancients, we have not only our eyes but also telescopes. What does a telescope do? Before you answer with a chorus of "It magnifies" or "It brings things closer," let me tell you that you are quite wrong. As far as stars are concerned, a telescope neither enlarges nor brings them closer. Sorry to disappoint you, but I have to remind you that the stars are exceedingly far away, and even a telescope with a 1,000-fold "magnification" cannot change this fact. A star is pointlike and remains pointlike even in the focal plane of a telescope. Why use a telescope in stellar observations, then? You must reflect a moment on how the eye works. Figure 30 will help. A star radiates light throughout space. (There is no reason to think, at least when we confront the problem in general terms, that its emission has preferential directions.) A fraction of this light reaches the eye—specifically, the light contained in a cone having the eye as base and the star's distance from the eye as height. Observe that it is a very narrow cone. Now, for the organ of sight (eye-nerves-brain) to work and for someone to *see*, the amount of light reaching that person's eye must be greater than a minimum value (or threshold value, which varies from person to person). If it is smaller, no matter how hard or how long you stare, you see nothing where the light source is supposed to be. A telescope receives more light than the eye because the area of the telescope's lens is larger than the area of the eye's pupil. In the dark the latter is about 20 mm^2; hence we can quickly calculate that a telescope 30 cm in diameter (a favorite with amateur astronomers) collects from the same star, and under the same conditions, about 3,500 times the light that reaches the eye; a 100-cm telescope (a medium size) collects 40,000 times more light. Since a telescope is built to convey to the eye all the light that it receives from a star, it follows that looking with a telescope is like looking with an eye having a pupil as large as the telescope's lens; with the 30-cm and 100-cm telescopes, we receive 3,500 and 40,000 times more light, respectively. Since a factor of 100 corresponds to 5 classes of magnitude, with the first telescope we can see stars up to magnitude 15 (if with the naked eye we see stars of magnitude 6), and with the second up to magnitude 17.5. In reality, telescopes are not quite so efficient; there are losses and other factors that reduce visibility by 1 or 2 magnitudes. But there is no question that stars that were once

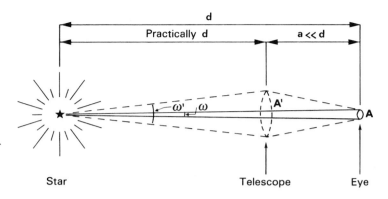

Figure 30
Diagram illustrating the use of the telescope in a stellar observation. The eye, at A, can collect the light contained in a cone whose vertex is the star, that is, the light contained in the solid angle ω. But the telescope can gather, at A', and convey to the eye the light contained in the solid angle ω'. Given the enormous distance from the star to the eye, the distance from the star to the telescope (*d* − *a*) is practically the same as that to the eye (*d*).

invisible become visible. There are myriads of such stars. If you point the telescope at regions of the Milky Way, you can hardly believe your eyes. And even if you point it at regions of the sky that seem empty, you see more stars than you can count.

It follows, then, that a telescope neither enlarges nor brings stars closer, but rather enables us to see celestial objects where we could have sworn there was nothing. Things change somewhat in the case of extended sources, but we shall not go into that because it would take us out of our way.

The original scale of six magnitudes has been extended to include the fainter stars that have been discovered, first with telescopes, and then by means of accessory instruments such as photographic emulsions, photoelectric cells, electron tubes, and image intensifiers—all of which increase the telescope's power almost to its theoretical limits (figure 31).

After choosing reference stars for each magnitude,[26] by comparison (with some precautions I shall not specify) we can establish the magnitudes of all the others. Apart from directly comparing two stars, it turns out that their apparent magnitudes may be determined by the following relation between them (called Pogson's relation):

$$m_1 - m_2 = -2.5 \log S_1/S_2,$$

where m_1 and m_2 are the magnitudes and S_1 and S_2 the "signals" (or, if you like, the light intensities) of the two stars.[27] Pogson's relation is the result of psychophysical studies carried out by Fechner. One can think of m as the sensation of an organ (sight in this case) and S as the strength of the stimulus (light in this case). Note that to have a difference of 5 on the left-hand side, we need a ratio of 100 between S_1 and S_2 on the right.

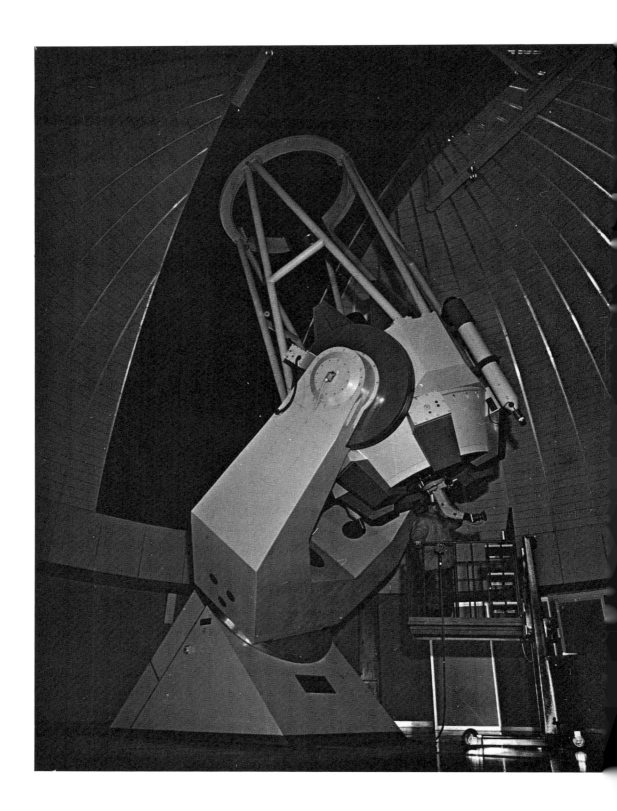

Figure 31
The 182-cm telescope of
the Astrophysical Observatory at Padova–Asiago. It is
the largest telescope in
Italy.

The magnitudes obtained in this manner are called apparent because they represent what *appears*, what we *see*. Although we usually say that star A *is* brighter than star B, the use of the verb *to be* is incorrect. All we can say, to be correct, is that star A seems, appears brighter than star B, just as a lighted match is (appears) brighter than, say, a burning forest 20 km away. Similarly, we cannot assume that a star that appears brighter is actually, intrinsically brighter. To make an accurate comparison that would justify the use of the verb to be, all the stars would have to be at the same distance. And they are not. But since we are quite resourceful and have worked our way out of tighter corners, we shall invent something.

If we take a lamp and push it farther and farther away from our observation post, we soon realize that it becomes fainter and fainter. The reason is easily explained. Light propagates in all directions and therefore spreads over an ever larger space. If a light source is at point O in figure 32, the observer at A receives at *a* in a given period of time an amount of light whose ratio to the total light emitted at O in the same period equals the ratio of area *a* to the surface area of the sphere of radius OA. For observer B at distance OB, the light that before spread over area *a* now spreads over the larger area *b*. Hence for B the light becomes fainter by the ratio *a/b* of the two areas. Since the area of a sphere is, as everyone knows, $4\pi r^2$, the ratio *a/b* is equal to the ratio of the corresponding radii squared, that is, $(OA)^2 / (OB)^2$. It follows that the amount of light that reaches a given area (say, a square centimeter) diminishes as the distance increases, or, more precisely, as the square of the distance. Consequently, the brightness of a star also diminishes by the square of its distance from the observer. Given two stars A and B of the same intrinsic luminosity, star A will appear four times brighter than B if B is twice as far away as A, and nine times brighter if B is three times as far away as A.[28]

Now we can ask, How bright would the stars appear if they were all brought to the same distance from us? We know the distances of some stars—not too many, but not too few either— and for these stars at least we shall be able to make direct comparisons of brightness. Choose a standard of distance. Then if we know that a given star must be "brought" 3 times closer so as to be at the standard distance, we shall have to increase its brightness by a factor of 9 (its *brightness*, not its *magnitude*; recall that a jump of 1 magnitude corresponds to a 2.5-fold increase or decrease in

Figure 32
The light emitted by the
source at O propagates into
space. The amount of light
received at a given area is
inversely proportional to the
square of the distance of
the area from the source.

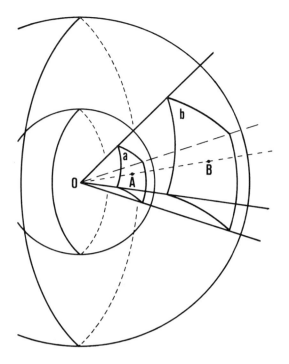

brightness). If in order to be at the standard distance a star must
be brought 10 times closer, its brightness must be increased 100
times. If a star instead must be moved 7 times farther away, its
brightness will have to be decreased 49 times. And so on. At the
end of the operation we shall be able to compare brightnesses.
Careful, now. Since the standard distance is wholly arbitrary, we
still do not know a star's intrinsic luminosity. But at least we have
defined a quantity that is directly related to the intrinsic luminosity.
This quantity gives a new magnitude, the so-called absolute magnitude, which is defined as the apparent magnitude a star would
have if it were at the standard distance. The standard distance
that has been chosen is 10 parsecs. With a few exceptions, therefore, all stars must be brought closer in order to obtain their
absolute magnitudes. If we call m the apparent magnitude, M the
absolute magnitude, and p the parallax of a star, a few calculations
produce the following relation:

$$M = m + 5 + 5 \log p.$$

That "log" (logarithm to the base 10) probably bothers you (I
know how much school has made us hate this sort of thing.) But
the fact is, without mathematics we cannot really study nature

(all of nature, not just astronomy). Given a star of known distance d or parallax p (it is the same thing, since $p = 1/d$, where p is in seconds of arc and d in parsecs), once we determine its apparent magnitude—something that can be done, roughly, even by eye— the relation just stated will give us its absolute magnitude. Naturally, what was said about the apparent magnitude applies to the absolute magnitude as well: The fainter the star, the larger its absolute magnitude. It is simply a convention.

The magnitude scale of the ancients has been extended on the minus side as well as on the plus side. To give some examples, the sun, which is very close, has an apparent visual magnitude of -26.74, Polaris of $+2.3$, Sirius of -1.45 (Can you tell how much brighter than Sirius the sun appears?) Vega of $+0.04$, and Aldebaran of $+0.85$. Every bright object can be given an apparent magnitude, including the moon, of course, and the planets. The apparent magnitudes of the latter, however, change all the time. (Why?) What are the absolute magnitudes of the stars mentioned? If we think of them at 10 parsecs, the sun will have an absolute visual magnitude of $+4.83$, Polaris of -4.6, Sirius of $+1.4$, Vega of $+0.5$, and Aldebaran of -0.7. If our sun, which does such a good job of lighting and warming the earth, were at 10 parsecs, we would hardly see it.

Although we set out to discover the actual amount of energy stars emit, we cannot do it yet. There are other things we must learn first, but we have certainly made some progress.

By measuring the distance of the few hundred stars that lend themselves to direct measurements, we have begun to get an idea of the size of the universe. Distances of 50 to 100 parsecs are already fairly large, but we know that there must be much larger ones yet, if for no other reason than that there are stars for which direct measurements cannot be made. We can also make direct comparisons between magnitudes, or, if you like, between brightnesses, at least for the nearest stars. Later on we shall use all of these things to make further progress.

·INTRODUCING THE QUESTION OF STELLAR TEMPERATURES

Meanwhile, let us see whether we can say something about the surface temperatures of stars. To begin with, they cannot be low. When heated, an object starts to shine. At a temperature of about

800°K it gives out a red light; the temperature of the lighted tip of a cigarette is of this order of magnitude. Heated to a higher temperature, the object becomes more luminous; the more luminous, the greater the temperature. At the same time, the color changes from red to yellow, then to white, then to a bluish white, and then to blue. The temperature of the filament in a 100-watt light bulb (found in every home) is of the order of 1,800°K, while the temperature of a voltaic arc is between 2,500°K and 3,000°K. What we have just observed is that there is a relation between the color of a light source and its temperature.

But what exactly does it mean that an object changes color as it becomes progressively hotter? Without getting worried, allow me to say a few words, the bare minimum, about the blackbody. It will help us to understand many things to come.

· THE BLACKBODY

Let us consider an enclosure thermally insulated from the rest of the world. Such an enclosure would make an ideal thermos; any hot or cold drink stored in it would remain hot or cold forever. Although we know how to make a thermos, it is by no means ideal; perfectly insulated enclosures do not exist and cannot be built. But since we are considering an ideal, we shall not worry about that. Inside the enclosure, at first things may be at different temperatures; but after a while everything in it—any objects within, the walls, and the interior space itself—will be at the same temperature, having reached what is called thermal equilibrium. Naturally, the cavity will be filled with radiation because everything radiates, whatever its temperature; the objects in it radiate, the walls radiate. Thus the interior space is full of radiation (cavity radiation), which falls onto the objects and walls and is then partly absorbed and partly reemitted by them. It is not too hard to understand that in conditions of equilibrium a relation must exist between the radiation absorbed by every square centimeter of surface and the radiation emitted by that surface. This relation has been found and states that for any given substance, *in conditions of equilibrium*, the ratio between the emission coefficient and the absorption coefficient depends solely on the wavelength of the radiation (λ) and on the temperature (T). This ratio is denoted $B(\lambda, T)$.

·DIGRESSION ON THE WAVELENGTH OF RADIATION

I doubt that I need to explain the concept of wavelength. Whoever reads this book probably knows it already. We speak of electromagnetic "waves" because many phenomena can be explained if radiation is assumed to propagate in the form of waves. Given a series of waves, we can determine the *wavelength*, which is the distance between the crests, or troughs, of two consecutive waves. Since radiation travels in a vacuum at 300,000 km/sec (denoted c; it is the speed of light), it follows that for a wave of length λ, $\lambda\nu = c$, where ν is the number of waves contained in c (figure 33). This number is the *frequency* of the wave (given in hertz); obviously, shorter waves have higher frequencies than longer waves.

The various kinds of radiation differ from one another in wavelength, and all the radiations between $\lambda = 0$ and $\lambda = \infty$ (infinity) constitute the electromagnetic spectrum. Starting from waves of very short wavelength, we have γ rays and x rays; and as the wavelength increases, we go to ultraviolet (uv) rays, visible light, infrared (ir) rays, microwaves, and radio waves. It is always the same phenomenon and it is only the wavelength that changes, just as it is only the height that changes when we say of people that they are short or tall, dwarfs or giants. Even though it is always the same phenomenon, we need different instruments to detect radiations of different wavelengths. Figure 34 illustrates the electromagnetic spectrum. The portion of the spectrum that is called "visible" comprises radiation that can be seen by the human eye. Radiation of wavelength around 4,000 Å[29] is visible and gives the sensation of a deep blue-violet; radiation of about 5,000 Å gives the sensation of green; radiation of about 5,900 Å gives the sensation of yellow; and radiation of about 6,600 Å gives the sensation of red. Radiation whose wavelength is longer than 7,200 Å is no longer visible. Scientists, however, have built devices that can detect radiation of any wavelength, such as photographic plates, photosensitive cells, microwave detectors, and radio receivers. Each of these instruments works only, or works best, in one region of the spectrum.

In the future, when we speak of light or radiation, we shall not speak of colors (which are matters pertaining strictly to human vision), unless we have a special reason for mentioning them, but rather of wavelengths and corresponding frequencies.

Figure 33
Illustration of the concepts of wavelength (λ) and frequency (ν).

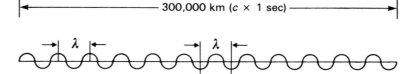

ν = number of waves contained in $c \times 1$ sec

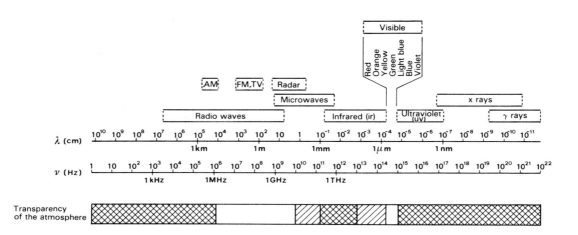

Figure 34
The electromagnetic spectrum, showing the various spectral regions. The AM, FM, and TV regions are part of the radio-frequency range. In the strip indicating the transparency of the atmosphere, the white areas mark the radio and optical windows, that is, the regions of the electromagnetic spectrum absorbed almost not at all by the atmosphere; the lightly shaded areas, the regions absorbed partially; and the heavily shaded areas, the regions absorbed totally. For example, the ultraviolet (uv) radiation, x rays, and γ rays coming from the sun are completely absorbed by the atmosphere. The units of measure often used for λ and ν are indicated; for unit abbreviations see the appendix.

•BACK TO THE BLACKBODY

Getting back to our enclosure, if we wish to observe the radiation in its interior, we have a problem. As soon as we make a hole in it to let the radiation through, we lose the conditions of equilibrium that we had achieved by thermally insulating it from the outside world. Hence the radiation that escapes is no longer what we wish to study. One way to solve the problem is to make a very large enclosure and, once equilibrium has been reached, a very small hole—just big enough for us to see inside without altering the conditions of equilibrium in any appreciable (measurable) way. The results of our observations are illustrated in figure 35, which needs a few words of explanation. In this graph each of these curves (known as Planck's curves) indicates the relative energies radiated at a given temperature at different wavelengths. Notice that the emission is not equally intense at all wavelengths; for each temperature, the emission, starting from $\lambda = 0$, increases as the wavelength gets longer, reaches a maximum at a certain wavelength, and then declines as the wavelength gets even longer. Furthermore, there is an overall increase in emission as the temperature rises, and an overall decrease as the temperature falls. Last, maximum emission shifts toward shorter wavelengths as the temperature rises. Let us now consider just one of the curves in the graph. For each wavelength, the segment between the λ axis and the curve represents the energy emitted per second by a square centimeter of the radiating surface at the given temperature (T). The sum of all the segments corresponding to the individual wavelengths obviously gives the area between the curve and the λ axis. Hence this area is a measure of the total energy (that is, at all wavelengths) radiated each second into half of space by a square centimeter of the emitting surface at temperature T. This quantity is called *emissive power*.[30]

Recall now that in conditions of equilibrium the ratio of the emission coefficient to the absorption coefficient is a function of T and λ, the same for all substances—or, as we say, a universal function of T and λ. In particular, since this is true for each wavelength, the ratio may be given in terms of the monochromatic (that is, single-wavelength) emission and absorption coefficients. The function denoting this ratio, which, as I said, is $B(\lambda, T)$, expresses Planck's curve. Given the temperature T (that of equilibrium) the value of B is fixed for each λ, but changes with λ. Conversely, we need only have one value of B, at a certain λ, to

Figure 35
Blackbody spectra at three
different temperatures. The
black dots indicate the continuous spectrum of the
sun's photosphere.

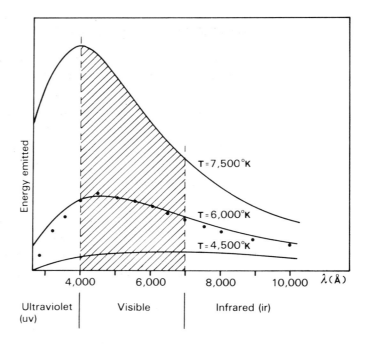

determine T, because only one of Planck's curves can have, for that T, that value of B at wavelength λ.

To put things in mathematical terms (a bad habit with us), the fact that under conditions of equilibrium the ratio at wavelength λ of the emissivity (monochromatic emission coefficient) to the absorptivity (monochromatic absorption coefficient) is $B(\lambda, T)$ can be written

$$\frac{\varepsilon_\lambda}{\kappa_\lambda} = B(\lambda, T) \qquad \text{or} \qquad \varepsilon_\lambda = \kappa_\lambda B(\lambda, T);$$

the second relation emphasizes that the emission and absorption coefficients, ε_λ and κ_λ, respectively, are those for a specific λ.

It may be useful to add that these relations hold not only for coefficients relative to a unit of area but also for those relative to a unit of volume or mass.

Now let us consider a body that can absorb all the radiation that falls upon it. For this body, which does not exist, $\kappa_\lambda = 1$ for all the values of λ.[31] It is called a *blackbody*. Do not be deceived by the name; the blackbody is not black. On the contrary, it is intrinsically brighter than any other body at the same temperature. As we say, it is the perfect radiator. Let me explain. In conditions of equilibrium, the emission coefficient of a body is $B(\lambda, T)$ times a certain κ_λ, which is surely less than 1; for any body, therefore,

ε_λ is smaller than B_λ. For the blackbody, instead, $\varepsilon_\lambda = B_\lambda(T)$,[32] which is the maximum value an emission coefficient can have at wavelength λ and temperature T.

The blackbody *appears* black at room temperature. At this temperature objects radiate practically in the infrared, which we cannot see, and therefore appear black. But between a blackboard and a bed sheet, the latter would seem to radiate more; after all, it is white, while the blackboard is black. Wrong. The sheet radiates less. Its whiteness is due to sunlight that hits it and is reflected, not to intrinsic radiation emitted by it at that temperature. The blackboard absorbs almost all and reflects almost nothing. If we could observe both objects in the infrared in a dark room, so as to reveal the intrinsic emission of each body, we would see the blackboard as much brighter than the sheet. Thus as the temperature increases, a black (or almost black) body keeps on radiating more than any other body; at temperatures of the order of 5,000°K to 6,000°K, the blackbody shines like the sun. As the blackbody becomes hotter and hotter, it changes color; from a deep red it turns to bright red, to yellow, to bluish white, to blue. This is because even though emission increases at all wavelengths, the emission peak shifts toward increasingly shorter wavelengths, that is to say, from the red toward the violet.

I hope I have not bored you, but I have gone through all this for a purpose. First, to make you understand how many phenomena must be taken into account and how much reasoning is needed to acquire a knowledge of the sky (and you will soon see the consequences of what you have just learned). Second, to make it plain that astronomical research entails a bit more than enjoying the beauty of the starry sky. And last, to stress that when we make a scientific statement, we must always define its terms and therefore its degree of validity or approximation.

·THE COLORS AND THE TEMPERATURES OF THE STARS

Let us return to the stars. Each star has a color. There are red stars, yellow stars, and blue stars. Even with the naked eye it is easy to see that Aldebaran is red, Arcturus is orange, and Vega is white. But the multitude of fainter stars that dot the sky show no obvious coloration. This is due to the nature of the human eye, which at low levels of luminosity loses the power to distinguish colors. And the stars look indeed very faint. To see the color of a star we must use cameras and special photographic techniques. We can take color pictures, but we can also use black-and-white film first with a red filter and then with a blue one. If a star is blue, it is because it emits mostly in the blue. From the star, therefore, we shall receive much more blue light than red light. The red filter lets the red light through, whereas it blocks the blue light. The blue filter lets the blue light through and blocks the red light. The photograph of the blue star taken with the red filter will give a faint stellar image, while the photograph taken with the blue filter will give a brighter image—a sure sign that the star is intrinsically blue. By an analogous procedure we can recognize red stars.

The color sensitivity of the human eye is greatest in the yellow-green. Thus a star that emits mostly in this region will look brighter to the eye than a star that, at an equal distance and emitting an equal amount of energy, has maximum emission in other spectral regions. (That is, given two equally intense luminous signals, one yellow-green and the other of a different color, the eye sees the former better.) From the observed brightness, if you recall, we obtain the apparent magnitude. In addition to the apparent *visual* magnitude, we can now determine the apparent *photographic* magnitude (with or without filters). If we use normal photographic emulsion, which, unlike the eye, is most sensitive to violet, we obtain a photographic magnitude that we can compare with the visual. Thus for a given star we can find the number

$$C = m_{\text{photo}} - m_{\text{visual}},$$

which is called color index and by convention is set to 0 for certain white stars such as Vega. Since a blue star appears brighter to the "eyes" of the photographic plate than to the human eye (the photographic plate is more sensitive to blue than our eyes), its apparent photographic magnitude must be smaller than its apparent visual magnitude; hence the color index of a blue star is negative. Conversely, a red star has a positive color index.

a temperature index; the higher the color index (on the plus side), the lower the surface temperature of the star. We still do not know what the actual temperatures are, but we can now rank the stars in order of temperature.

Let us make a bold hypothesis. What if the stars were blackbodies? It seems out of the question. We said earlier that to observe blackbody radiation we have to do so without altering its conditions of equilibrium—for example, by observing it through a small hole in the body. Then how can we consider a star as a blackbody when it is losing energy on all sides? The very fact that we can see it tells us that it cannot be a blackbody. True. But a few calculations, which we will not make at this time, show that the energy emitted is but a small fraction of the energy contained in the star. Thus we can make the assumption, admittedly not quite right, that to a first approximation stars radiate as blackbodies. To a first approximation, I said, and in fact stellar physics does not stop here. Nevertheless, the results we obtain, though rough, are not very far from the truth, and all subsequent refinements of the blackbody model leave the basic assumption unchanged. By assuming that stars are blackbodies we can begin to estimate stellar temperatures.

To do this, we use the color indexes. Suppose we find that a given star's color index is $+2$. Next we turn to the family of Planckian curves and identify the curve (the only one) for which the ratio of the intensities in the yellow-green and in the blue gives the color index $+2$. Once we have found the curve, we have also found the temperature because for every temperature there is a very precise curve. Now we can put numbers in our scale of stellar temperatures—another big step forward. These numbers are a little rough, due to the roughness of the method, but they give a pretty good indication of the actual values.

Surface temperatures range from a low of about 2,000°K to a high of 50,000°K, and we shall shortly find confirmation that our results are correct.

· THE ORIGIN OF RADIATION

To learn more about the stars, we must first talk about stellar spectra; and before that, about spectra in general; and before that, getting down to fundamentals, about the emission of radiation by matter. The subject is extremely complex, and I hope that, once again, you will forgive the inevitable imprecisions and simplifications.

I am sure you have heard that matter consists of atoms and that atoms, in turn, consist of various components, the most important of which are protons, electrons, and neutrons. Although we could proceed without discussing the structure of the atom, it might be a good idea to say a few words on the subject, particularly because it will be useful later on.

There are various models of the atom, and I want to make sure you understand that all such models have been advanced to describe, not so much how the atom is made, but rather how it works, how it behaves. This is an aspect of scientific research that is never stressed enough, not even by schoolteachers, let alone by popularizers.

A model that holds for all the phenomena so far known is put forward as a working hypothesis; it then can be perfected, simplified, extended, changed, thrown out, or added to at need. When we say of the atom, or any other thing, that it is "made" in such and such a way, what we mean is that as far as we know, in its interactions with other things the atom behaves *as if* it was made in such and such a way. Whether it is *actually* made that way is an entirely different matter; it is a problem for philosophers, not scientists. Doing science today means studying phenomena, that is, interactions between entities, which, in the final analysis, can be intrinsically anything they please. For an entity to have any meaning, it must be involved in interactions with at least one other entity. Consider a person who is isolated from all other people and unable to interact with anything at all. Even assuming that we know this person exists, what can we say about such a person? That the person is good? That the person is tall? Every question assumes a comparison or some sort of interaction with the world, and the world of people first of all. It is the interactions, the relations, with which we are concerned.

Hoping to have made my point, I return to the atom. Among current atomic models, the simplest (but more than adequate for our purposes) is the so-called planetary model. In this view the atom is pictured as a microscopic planetary system with the nucleus at the center and one or more electrons revolving about it in circular orbits. The nucleus consists of protons and neutrons and contains, practically speaking, the entire atomic mass. Each proton carries a positive elementary electric charge. A neutron has the same mass as a proton and can be considered as a proton devoid of electric charge. Under normal conditions, the nucleus is surrounded by as many electrons as there are protons in the nucleus. Since each electron carries a negative elementary electrical charge, the atom as a whole is electrically neutral. All ellectrons are alike, and each has a mass about 2,000 times smaller than the mass of the proton; hence their total mass is negligible compared with the mass of the nucleus. The simplest atom, the fundamental constituent of the universe, is the hydrogen atom, which has only one proton in the nucleus and one outer electron. Couldn't be simpler. Next comes helium, with 2 protons and 2 neutrons in the nucleus and 2 outer electrons. Then come lithium (3 electrons), beryllium (4 electrons) and, one by one, all the natural elements up to uranium (92 electrons).

Although the planetary model is often likened to a miniature solar system, the analogy is not quite correct. Unlike the solar system, in which the planets always travel along the same orbit, in the planetary model the electrons keep on jumping wildly from one "orbit" to another. (The "orbits" are not selected at random, however, but follow precise rules.) The bond between the nucleus and the electron (electrostatic attraction) is stronger when the electron is in one of the inner orbits than when it is in one of the outer orbits. Thus if an electron in an inner orbit is to be made to move to an outer orbit, it must be supplied with some energy to overcome the attraction of the nucleus. Consider a pendulum at rest. For the pendulum to move, a certain amount of work must be done and something must do it, let us say your hand. By moving the pendulum, your hand does the necessary work, and this work goes into increasing the (potential) energy of the pendulum. Released, the pendulum starts to swing and does work (it warms the air, it breaks a glass in its path, and so on) until all the energy it had stored is expended, whereupon it comes to rest again. In the context of the planetary model, the electron behaves in approximately the same way. Excited by a collision or by the absorption of energy by the atom, the electron jumps to an outer orbit, where it remains for a very short time (as if this outer orbit were not its natural position); then it jumps back to an inner orbit and eventually, unless reexcited in some manner, to the innermost orbit—where it will not rest for very long if surrounded by many other atoms or radiation that can excite it.

Moving from an inner to an outer orbit, the electron absorbs energy. On the return trip it loses energy, and this energy is emitted as radiation. The larger the jump, the greater the amount of energy involved. When energy is lost, it is lost in one lump sum, as a packet of energy, a tiny ball of energy, a corpuscle of energy, a burst of energy. And it leaves the atom as a pulse of radiation, a corpuscle of radiation, a whatever-you-like of radiation. It is called a *photon* (or a *quantum* of light) and it behaves in a rather odd way. Sometimes it interacts with matter as if it is a particle (it can be absorbed by some atoms or molecules; it can be deviated from its path; it can rebound; and it can penetrate matter like a bullet) and sometimes as if it is a packet of waves (it can produce phenomena of interference, even with itself).

The energy carried by a photon travels through space as radiation. Recall that to radiation is associated a wave of a certain length. There is a relation between the energy of the photon and the wavelength of its radiation; it is given by this very simple formula,

$$E = \frac{hc}{\lambda},$$

where E is the energy, h a numerical constant (Planck's constant), c the speed of light, and λ the wavelength. It follows that the larger the jump made by the electron inside the atom, that is, the larger the energy emitted (or absorbed), the shorter the wavelength of the radiation. Photons of high energy (short wavelength) give the eye the sensation of violet, while photons of low energy (long wavelength) produce the sensation of red. Photons of higher energy than that corresponding to the extreme visible violet can be detected with ultraviolet, x-ray, and γ-ray techniques. Photons of lower energy than that corresponding to the extreme visible red can be detected with infrared, microwave, and radio techniques.

There is another fact to be considered. Atomic structure varies from atom to atom, so that the possible jumps of electrons within atoms (hence the wavelengths of the radiations an atom can absorb or emit) are different for different atoms. This has an interesting consequence, namely, the possibility that through an analysis of the radiations emitted by a body, we may be able to do a chemical analysis of that body. Suppose we know that a certain element or compound can only emit specific radiations and suppose that these radiations are found in the light emitted by a body whose chemical composition we do not know. It would not be unreasonable to assume that that element or compound is present in the body in question. We can certainly experiment with all available materials to make sure that we are on the right track. If it works, this method will furnish further proof that our atomic model is correct. Moreover, it will give us the key to understanding what the stars are made of without ever actually going to them. If we are right, the ray of light from a star is its identity paper, and on that ray of light is inscribed a message. And therefore, since we cannot go to the stars, we must learn to read as much of their message as we possibly can. But to read a message written in code, we must have the key to the code. Thus our next task will be to talk (briefly, I promise) about spectra. You have already learned something about them, but a little more is needed if we are to make quick progress. Before we start, I would like to change a word that I have used in my description of the planetary model to indicate the places occupied by the electrons. Instead of "orbits," from now on I shall speak of "energy levels." Since the various positions of an electron in relation to the nucleus correspond to different energies of the atom, the term energy levels is more convenient; it is also more correct, in that it is not tied to any particular atomic model. I have said that the attraction between the nucleus and the electron is strongest when the electron is in the innermost orbit and is weaker when the electron is in one of the outer orbits (the atom having acquired energy in some manner to make this transfer possible). Thus instead of speaking of jumps of the electron between orbits, we can speak of jumps between energy levels. Energy is absorbed in the transition from a lower to a higher level, and energy is released in the transition from a higher to a lower level. We can make a schematic representation of the energy levels (called an energy-level diagram; see figure 36) by drawing horizontal lines at intervals proportional to the differences in energy between succeeding energy levels. Every level (also called an excitation level) represents the energy necessary to bring the electron from a lower level to the level in question. The atom, however, cannot be excited beyond a certain limit; sooner or later something must give. For a certain value of the excitation energy, characteristic of every atomic configuration, the bond between the nucleus and the electron breaks and the electron escapes from the atom. What is left in the atom (now called an ion) are the nucleus and possibly other electrons. In this state the atom is said to be ionized. Bereft of an electron, the atom is still an atom of the same kind; if it was an atom of iron, it still is an atom of iron, but it is an ionized atom of iron. If it has lost one electron, it is a once-ionized iron atom; if it has lost two electrons, it is a twice-ionized iron atom; and so forth. For every atom, however, a different state of ionization corresponds to a different configuration of the energy-level diagram; ionized atoms behave differently from nonionized atoms, and in this sense they may be regarded as different atoms.

In the energy-level diagram we can now plot a bottom (horizontal) line, corresponding to the fundamental level (also called the ground level or level 1), and a top (horizontal) line, corresponding to the ionization energy. How many levels can there be between the

Figure 36
Energy-level diagram for the hydrogen atom. Transitions from the second level to the upper levels produce the lines of the Balmer series by absorption, while transitions from the upper levels to the second produce the same series by emission. The Balmer series spans the visible range. The transition 2 → 3 (that is, from the second to the third energy level) produces a line (Hα) at λ = 6,562.8 Å; the transition 2 → 4 produces a line (Hβ) at 4,861.3 Å; the limit of the series, corresponding to the transition 2 → ∞, is at 3,650 Å, in the near-ultraviolet. The figure shows the α and β lines of the first three series—Lyman (far-ultraviolet), Balmer, and Paschen (infrared)—and the α line of the next three series—Brackett, Pfund, and Humphreys (all in the far infrared).

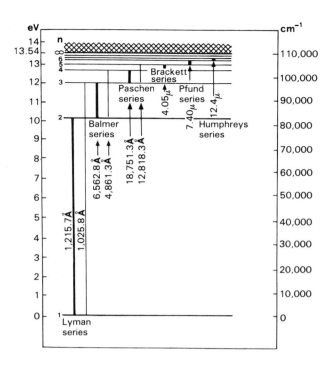

two lines? The answer is, an infinite number. This is not at all in contradiction with our previous statement that the orbits of the electrons are not random but determined by specific rules. Suppose we draw a bar and mark on it the midway point. Next we mark the midway point on its right half. Then we mark the midway point on its right quarter. And so on. We could go on marking points for the rest of our lives and pass on this delightful exercise to our children, grandchildren, and all subsequent generations until the end of time without ever being able to say, There, this is the last point, I have reached the end of the bar. It is the same with the atom's energy levels. They are discrete, that is, distinct, but infinite in number and are contained in the band of energy comprised between the fundamental level and the ionization level. The levels are well spaced at first, but they soon crowd together, much like the points on our bar, but according to more complicated rules. Figure 36, which illustrates the energy-level diagram for hydrogen, happens to be very simple; but such diagrams become fairly complicated for more complex atoms. For our purposes, however, we do not need to complicate our lives further. Life is difficult enough as it is.

Suppose now that we have a certain amount of gas (I have chosen gas because it is a relatively simple physical system) and that the gas is hot enough for the atoms to become excited (by colliding with other particles or by absorbing radiating energy). Suppose further that the gas consists entirely of hydrogen atoms (once again, I have chosen the simplest case; conceptually, nothing is lost). As we have seen, the excitation of the atoms causes the electrons to jump from lower to higher levels. The atoms, however, remain in an excited state only a very short time (roughly one hundred-millionth of a second) and, unless reexcited, the electrons fall back to lower levels, releasing the previously absorbed energy as radiation. Let us follow the fate of an electron at ground level. It gets excited, jumps to a higher energy level, and immediately falls back, either in a single jump (back to ground level) or in a number of jumps. This corresponds to the emission of either a single photon of energy E or a number of photons of energies E_1, E_2, and so on, whose sum is E. (Recall that each photon has a wavelength that is inversely proportional to its energy: the higher the energy of the photon, the shorter the wavelength of the emitted radiation.) The same happens in the case of electrons that are already at higher levels. For example, an electron that is at the fourth level can jump to, say, the seventh before falling back to lower levels. From the seventh the electron returns to the first, or ground, level because that is where it tends to be, but it can get there in one or more jumps. In each jump it emits a photon whose energy is equal to the difference between the energies of the two levels.

How many jumps can an electron make? We have seen that although discrete, the energy levels are infinite in number; consequently, the jumps, as well as the emitted radiations, are also infinite in number. In reality, the transitions between higher levels are not very different; they produce radiations whose wavelengths differ so little that from a certain point on they become, first very hard, and then impossible to distinguish, or rather separate, experimentally.

Although the number of transitions is theoretically infinite, in reality we have only considered transitions between discrete levels, which are termed *bound-bound* transitions, in that they occur between specific energy levels. But there are others. As we have seen, if the energy absorbed is sufficiently large, the electron can escape from the atom. The electron expends part of the energy in the ionization process and keeps the remainder as kinetic energy. Thus the electron leaves the atom with a certain velocity, which is higher the larger the energy left to the electron after the ionization process. This can be illustrated in our energy-level diagram by saying that the electron can occupy any level above the ionization level. This means that we have extended the diagram above the ionization level. We can think of the ion and the electron as still being a single system, whose energy levels are the discrete levels of the ion with an extra electron and a continuum of levels above the ionization level of the ion. When an atom is ionized, we can thus say that the electron jumps from a discrete level to a level of the continuum; this is called a *bound-free* transition. Conversely, when an ion captures a free electron, the electron must end in one of the possible discrete levels of the atom (which becomes neutral again if it lacked a single electron, singly ionized if it lacked two electrons, and so on); we can thus say that the electron has jumped from a level of the continuum to one of the discrete levels and call the jump a *free-bound* transition, which corresponds of course to the emission of a photon of energy equivalent to the jump. A third type of transition, termed *free-free*, corresponds to electron jumps between levels of the continuum. These occur when the electron, passing close to ions, changes its velocity, that is, its energy, with an absorption or loss of energy. The lost energy is emitted as a photon of equal energy.

All I have said, complicated as it is, is still a fairly simplified picture. For example, a downward jump may occur without emission of the corresponding photon because an atom or electron can transfer energy to other particles by accelerating them. We can disregard these special cases, which do not have a preeminent role in stellar atmospheres owing to the thinness of the gas, and keep well in mind the main phenomenon—deexcitation by emission of radiation.

Getting back to our gas, suppose that by a special cinematographic technique we can observe what happens to its individual atoms. As the film starts rolling we see electrons jumping upward and downward all over the place. Of course, each atom may be doing different things because the behavior of an atom depends to a large extent on what is happening in its vicinity. Let us now stop the projector on a single frame. Given a sufficient number of atoms, we shall see the electrons of the various atoms distributed at all possible energy levels. And the atoms are always in sufficient numbers because, rarefied as they may be, stellar atmospheres have densities on the order of a hundred or a thousand billion atoms per cubic centimeter. Were they a hundred times less dense, they would still have densities on the order of a billion atoms per cubic centimeter. Thus the number of atoms is very large even in a thin gas.[33]

Let us roll the film again. We shall see some electrons move to lower levels, emitting photons while doing so, and others move to higher levels, making all the possible upward and downward jumps. In other words, at any instant (given the large number of atoms and bearing in mind that an atmosphere consists of much more than 1 cubic centimeter of gas) our gas emits all the possible radiations that can be obtained from the corresponding energy-level diagram, that is, radiations due to "bound-bound," "free-bound," and "free-free" transitions.

There is one more thing to discuss. As we have seen, there are two mechanisms whereby an atom can absorb energy, namely, by collision with other particles and through the capture of a photon. This raises a question: Which photons can be absorbed by and excite the atom? The answer, roughly speaking, is the photons that have an energy, and hence a wavelength, corresponding to one of the possible jumps in the energy-level diagram. Photons having energies other than these pass the atom undisturbed, as if they did not exist. Think of a cigarette machine. If you put in the right coins, the machine "absorbs" them and immediately "emits" something of equal value (cigarettes). If you put in the wrong coins, the machine does not absorb them and, unless broken, gives them back.

In a hot gas atoms and particles are in thermal motion; the inevitable collisions cause transfers of energy, excitation of the atoms, and emission of radiation. Under these conditions a radiation field forms. If the gas is not enclosed in an opaque cavity, all or part of the radiation leaves the gas and spreads through space.

It is the same with a stellar atmosphere, with one major difference; its range of emitted radiations is all but infinite (because a star's atmosphere contains more than one type of gas), and each gas has its own energy-level diagram, which is generally very complicated. Each gas, moreover, may be neutral, ionized once, or ionized many times.

How can we make sense of what we see? The importance of analyzing starlight is now more than obvious. If the radiations emitted by a star are closely tied to the conditions of its atmosphere, those radiations can tell us all we want to know about the atmosphere—its temperature, for one thing, because the conditions of excitation change with temperature, so that the associated emission does too. At low temperatures, for example, the largest jumps are not possible and certain atoms cannot be ionized. A star's radiation can also tell us something about the density of the star's atmosphere (because if the density changes, the number of collisions changes) and about its chemical composition (because for a given chemical composition, certain radiations will be emitted and not others).

·STELLAR SPECTRA

The spectrum of a light source is the sum of the radiations emitted by the source. These reach the observer in the same instant because all of them, regardless of their wavelengths, travel through space with the same velocity (the speed of light). To determine the spectrum of a light source is to find out how its flux of radiation is distributed among the wavelengths of which it is composed. In practice, this is done by gathering a beam of light from the source and decomposing it into its constituent radiations.

Suppose we have a mound of bread crumbs of different sizes, ranging from tiny grains to much bigger pieces of bread. We might wish to determine the "spectrum" of sizes of our bread crumbs. To do this, we put the crumbs in a very fine sieve that lets through all the pieces smaller than, say, 0.02 mm. Then, setting aside this first mound, we pass the remainder through a slightly coarser sieve, which will let through pieces from 0.02 mm to 0.04 mm. Setting aside the second mound, we put what is left in a sieve that lets through pieces from 0.04 mm to 0.06 mm. We go on this way until the original mound is all gone and we are left with a number of small mounds, each containing bread crumbs whose sizes are known to within 0.02 mm. Finally, having weighed the various mounds with a very accurate scale, we make a graph by plotting the weights of the mounds (expressed as the number of crumbs in each of the mounds)[34] on the ordinate (vertical axis) and the crumb sizes on the abscissa (horizontal axis); see figure 37. This graph can quite properly be called a spectrum of the bread crumbs. For each value of the abscissa, the ordinate gives the corresponding weight; to put it another way, for each crumb size on the abscissa, the ordinate gives the relative contribution of crumbs of this size to the total weight (represented by the area comprised between the curve and the abscissa). This curve can be called a spectral distribution by weight of the bread crumbs as a function of bread-crumb size. A quick look at the graph tells

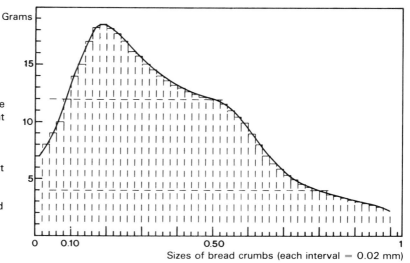

Figure 37
Spectrum of a mound of bread crumbs. Naturally, the spectrum would be different for a different mound of bread crumbs, one obtained, for example, from the same piece of bread but with a different grater, or with the same grater but from another piece of bread (drier, moister), and so forth.

us that there are three times more 0.51 cumbs than 0.81 crumbs.

The procedure is analogous in the case of radiation. What we do is plot the spectral distribution of the energy emitted by a light source (stellar atmosphere) as a function of wavelength. We already know the spectrum of the blackbody; it is none other than Planck's curve, which tells us the energy radiated at each wavelength by a blackbody at temperature T. To find out the spectral distribution of the energy emitted by a star, we must analyze its light. This is done with an instrument called a spectroscope when the image is viewed with an eyepiece, and a spectrograph when the eyepiece is replaced by a photographic plate or other recording device.

The spectrograph (figure 38) consists essentially of a closed box with a narrow slit, a few hundredths of a millimeter by as much as a few centimeters, that lets the radiation through. The optics inside the box does not see the light source but only the slit illuminated by the source. For the observer, therefore, the slit is the light source; but it is obvious that by analyzing the light from the slit, the observer is in fact analyzing the light from the actual source. First, the light source is focused by telescope on the slit. Then, after passing through the slit, the diverging beam is collimated (made parallel) by a lens, called a collimator, and them directed through a dispersing medium, which in its simplest form is a glass prism. This is the most important piece of the spectrograph; it separates the beam of light into radiations of different wavelengths, which emerge from the prism as distinct beams. The prism has this effect because in passing from the air to the glass of the prism (and then from the glass of the prism back to the air) light is refracted, or bent, by an amount that depends on its wavelength (is a function of the wavelength): the shorter the wavelength of the light, the more it is bent. Finally, a camera lens gathers the radiations dispersed by the prism and brings them to a focus on the focal plane, where they are still separate because radiations of different wavelengths arrive at the camera lens from different directions. Since the light source is a slit, the collimator and camera lenses, by themselves, would essentially give *one* image of the slit, that is, a line. Since there is a prism, however, the result is a series of lines dispersed (aligned) on the focal plane in order of wavelength. If the ray of light entering the slit contained only the radiations of hydrogen, on the focal plane we would have the spectrum of hydrogen; if the light source contained only calcium, we would have the spectrum of calcium; and so on. But if the source contains various elements, the spectrum we see consists of the spectra of the individual elements, all mixed together because the combined spectrum is arranged according to wavelength. Figure 39 shows the individual spectra of hydrogen and mercury as well as their combined spectrum. When we observe by eye or use color film we see a colored spectrum; but if we use black-and-white film, as is normally done, the result is an irregular series of lines that the observer must interpret by assigning the individual lines to the various elements that emit them. It is not an easy job, but at the moment we are not concerned with the difficulties involved. Basically, the wavelength of each line must be measured, and then the spectrum so determined must be compared with the known spectra produced by the gases of each element. This is not all there is to it, but we do not need to go into the details of analytical techniques. Suffice it to say that they enable us to identify the elements contained in a source.

By identifying the spectral lines we obtain a qualitative chemical analysis of a stellar atmosphere. To make it quantitative we must take into account the models of atmospheres and the mechanisms by which atmospheres emit radiation. This done, we can finally say which elements are found in stellar atmospheres and in what proportions. Before we start interpreting stellar spectra, however, there is something else we need to understand.

If the emitted lines are right next to each other, that is, if the source emits radiation at all wavelengths, we have a continuous spectrum. Such is the spectrum of the blackbody.

Figure 38
Diagram of a prism spectrograph. At the focal plane we observe a monochromatic image of the slit for each wavelength of the incoming radiation. For the sake of simplicity, the beam of light from the source is assumed to be composed of rays of two wavelengths, λ_1 and λ_2. Since rays of shorter wavelength are deflected more by a glass prism, $\lambda_1 > \lambda_2$.

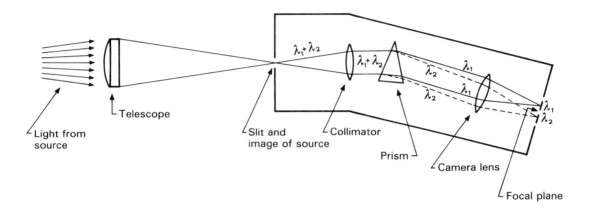

Figure 39
Diagram (in the visible spectral region) of the individual spectra of hydrogen (H) and mercury (Hg) and of their combined spectrum observed when the light source contains both elements (H + Hg). The spectrum of a star results from the superposition of the spectra of all the elements found in its atmosphere.

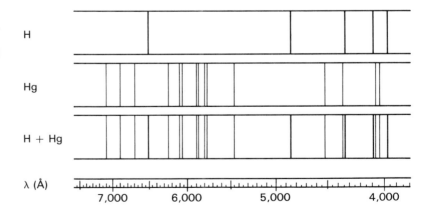

Let us consider now a rarefied gas. Essentially such a gas emits a spectrum of discrete thin lines, corresponding to bound-bound transitions, as well as a (more or less weak) continuum corresponding to free-bound and free-free transitions. If the temperature remains the same but the density increases, the lines become wider—the wider, the higher the density. There is no simple and immediate explanation for the broadening of spectral lines because this effect is due, not to one, but to several factors of varying importance. Not to leave you with an unsubstantiated statement, however, I shall try to describe how density can affect the width of the lines.

The fact that two particles come close to each other (atom to atom, ion to ion, electron to atom or ion, ion to atom) has certain consequences for the energy levels of the particles involved. Each "feels" the other, and the more so the closer they come. What we mean by "collision" is actually the passing of two particles at a very short distance from each other, that is, at a distance such that their energy levels are altered. The acting forces here are electrical in nature; ions and electrons are electric charges, and each creates around itself a field of influence called an electric field. I am sure you know that electric charges repel or attract each other depending on whether they have the same or opposite sign. The forces acting between nearby particles are of this type.

Suppose an atom is perturbed by a close encounter with a passing ion. The atom's energy levels are suddenly altered, causing the wavelength associated with any transition that happens to be taking place at that instant to depart from the value it would have were the atom unperturbed.

Let us see now what happens in a gaseous mass at a certain temperature and density. It makes no difference whether it is a boxful of gas or a stellar atmosphere. First, since the gas is hot, its component particles are in rapid and chaotic motion. This necessarily results in collisions between the particles. If the temperatue remains constant but the density increases, there will be a greater number of collisions. This happens not only for the case of atoms, but for all bodies moving at random; a battalion of soldiers marching in strict order is a group of individuals moving without collision, while people milling around in a small place cannot avoid collisions—unless they constantly correct their movements, which is something atoms do not do. Second, every atom in the gas undergoes, not just one but a succession of collisions, the frequency of which will depend on the density—the higher the density of the gas, the more frequent the collisions. Third, at every collision an atom's system of energy levels is altered. The result of all this is that the atom's energy-level diagram will not be made of horizontal lines but of horizontal strips, whose widths depend on the density of the gas. It is as if every level has vibrated about a position of rest corresponding to the position that the level has when there are no perturbations caused by other particles. Consequently, a transition between two levels, which originates a spectral line, does not produce a single radiation of well-specified wavelength, and hence a thin line, but rather a set of radiations that form a wider line. The higher the density of the gas, the wider the strips (energy levels) and the wider the spectral lines. Obviously, the role of excitation by collision is more important in dense than in rarefied gases.

It now remains for me to explain another feature of stellar spectra—dark lines on a bright continuum. To this end, let us make a laboratory experiment. First we place the glowing filament of an incandescent lamp in front of a spectroscope and analyze its spectrum. Like any other solid heated to red or white heat, the filament emits a bright continuous band of colors. Next we interpose a glass tube filled with a cool thin gas (sodium vapor, for example) between the hot filament and the spectroscope. We observe that the bright continuum from the incandescent lamp is now crossed by dark lines. If we then remove the lamp and excite the gas in the glass tube (by heating it or running electric current through it), we see a radically different spectrum. Instead of a brilliant continuum we see bright lines where the dark lines had previously appeared, while the places between the lines appear black.

Let us analyze this experiment. In the first place, it shows that the incandescent solid radiates a continuous spectrum. As the radiation emitted by the filament passes through the glass tube, it meets sodium atoms. This radiation consists of photons of all wavelengths, and hence of all energies, including those corresponding to the characteristic transitions of sodium. The photons of these specific energies are absorbed by the sodium atoms (although not necessarily all of them), while the others pass through the tube undisturbed. The absorbed energies are soon reemitted, as we know; but they are reemitted in all directions. As a result, the observer receives all of the radiation at the other wavelengths, but only part of the radiation at the characteristic wavelengths of sodium. At these wavelengths, therefore, we observe dark lines, known as absorption lines. A glowing gas, on the other hand, emits a characteristic bright-line spectrum, in which the bright lines, or emission lines, fall at precisely the same wavelengths at which the dark lines had been seen before.

·A SIMPLE MODEL OF THE STELLAR ATMOSPHERE

A stellar atmosphere radiates a continuous spectrum crossed by dark lines (of varying darknesses and widths) that fall at the same wavelengths at which we would observe the bright lines of the individual elements. If you recall our experiment, this spectrum is basically the same as that emitted by the incandescent filament placed behind a cool thin gas. This fact has suggested the following model of the stellar atmosphere. The deeper layers of the atmosphere, consisting of fairly dense gas, radiate a continuous spectrum. These layers, collectively known as the photosphere (light sphere), emit practically all of the radiation we see with the naked eye or the telescope, and if they were the only components of the atmosphere, a star's spectrum would be a continuum. Surrounding the photosphere, however, there is a layer of cooler gas, or reversing layer, that absorbs the radiation from the inner layers and produces dark lines on the continuous spectrum of the photosphere by the same mechanism by which the gas in our glass tube imprints dark lines on the continuum emitted by the filament. In figure 40 we reproduce a portion of the solar spectrum. Observe the continuum crossed by dark absorption lines, called Fraunhofer lines.

This stratified model is in reality an oversimplification. Since the model was first proposed we have attained a much better understanding of stellar atmospheres, and today we know that there is no reversing layer; both the continuous radiation and the absorption lines originate in the same region, namely, the photosphere. But it would serve no purpose here to go into the technical details of the current model. Furthermore, the old model is still valid for certain problems, at least to a first approximation. We can adopt it, therefore, provided we bear in mind that it does not correspond to reality and that a more sophisticated model must be invoked when accuracy is essential.

Within these limitations, let us proceed to analyze starlight. Saying "light" means restricting ourselves to the visible part of the spectrum. Of course, nothing prevents us from analyzing ra-

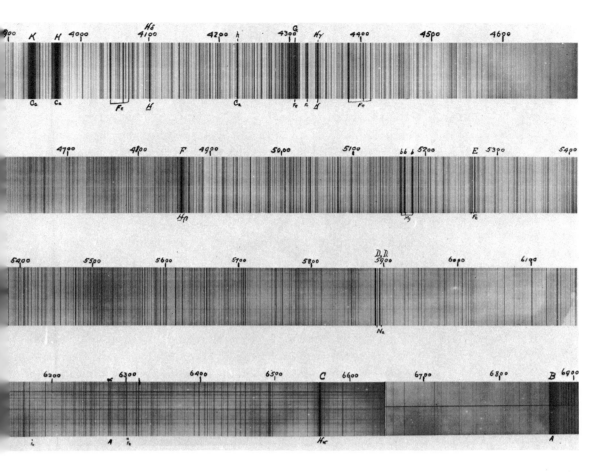

Figure 40
Visible region of the spectrum of the sun. A number of lines due to hydrogen (H), iron (Fe), magnesium (Mg), sodium (Na), and calcium (Ca) are indicated. Wavelengths are in angstroms.

diation of longer or shorter wavelengths; we simply have to employ instruments capable of detecting the different types of radiation.

To begin with, we shall disregard spectral lines and concentrate on the continuum. If you recall, we made the assumption that stars are blackbodies. We now have the means of evaluating the accuracy of our assumption. All we have to do is to compare the continuous spectrum of a star with the blackbody spectrum for various temperatures. Actually, the comparison is not wholly satisfactory; in general, it shows a fairly good fit in the visible part of the spectrum, but more or less marked deviations at the other wavelengths. Considering that we are dealing with an approximation, a perfect fit is not to be expected; but it is good enough to give us the order of magnitude of the quantities we seek.[35]

•STELLAR TEMPERATURES, LUMINOSITIES, AND RADII

A blackbody curve for a temperature of 6,000°K provides a good fit to the energy curve observed for the sun (figure 41). This means that if the sun were a blackbody, it would have a temperature of 6,000°K. But since the sun is not a blackbody, and since the blackbody is the body that for a given temperature radiates more than any other, the temperature of the sun's atmosphere must be somewhat higher, or it would not radiate as a blackbody at 6,000°K. Similarly, by finding the blackbody curves that best fit the observed energy curves, we can estimate the surface temperatures of all the stars whose spectra can be obtained.

An alternative method of temperature determination of a star, also based on the assumption that stars radiate like blackbodies,[36] consists in taking photographs with colored filters. For the wavelengths considered (which are determined by the filters used with the photographic emulsion) the intensities we find for the different colors are in ratios that are characteristic of a blackbody at a specific temperature. The temperature we find by this procedure is called *color temperature* for a given spectral interval because it is determined by an evaluation of color; in effect, we have compared the shape of the stellar spectrum with the shape of the blackbody spectrum, and the latter only tells us which color dominates in the chosen spectral interval. Since a star is not really a blackbody, the spectral distribution of the energy radiated by a star may differ in other spectral regions from that of the blackbody.

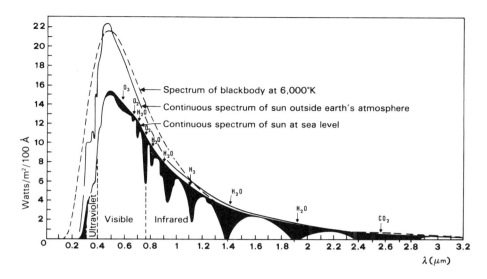

Figure 41
Comparison of the spectrum of the blackbody at 6,000°K with the continuous spectrum of the sun outside the earth's atmosphere. The bottom curve shows what is left of the solar spectrum at sea level because of absorption by atmospheric molecules. Indicated are ozone (O_3), oxygen (O_2), water H_2O), hydrogen (H_2), and carbon dioxide (CO_2).

There is another way of determining a star's temperature. Suppose we can measure the energy that we receive from a star and the star's distance. We can then calculate the star's luminosity, that is, the total amount of energy that the star radiates every second. In the case of the sun, for example, it has been calculated that a square centimeter of the earth's surface at right angles to the sun's rays receives energy equivalent to about 2 calories/ minute, or 1.36×10^6 ergs/sec. To determine the sun's total energy output per second, all we have to do is to multiply this solar constant by the surface area of the sphere (in square centimeters) whose radius is the sun-earth distance; that is,

$$L_\odot = 4\pi d^2 S,$$

where $L_\odot$ is the sun's luminosity, d is the earth's distance to the sun, and S is the solar constant.

Once we know a star's true luminosity,[37] we can replace it with a blackbody of exactly the same size and total energy output. How do we determine the temperature of such a body? All we have to do (in practice there are some difficulties) is find the Planckian curve that delimits an area numerically equivalent to the energy radiated by the star. Since there is an infinite number of Planckian curves, we shall certainly find the one we want. The temperature we obtain in this manner, which is generally different from the color temperature, is called *effective temperature*.

If stars were blackbodies, their color and effective temperatures would be the same. But a star is not a blackbody, and to the extent that these two temperatures are determined by different aspects of energy emission, they will be different—the more so the more a star differs from a blackbody. The effective temperature is the more satisfactory of the two because, first, it is determined uniquely and, second, in a way it expresses a star's total flux of energy. The color temperature can vary from one spectral region to another, and rather than a measure of the energy flux, it gives an indication of the shape of the continuum, that is, of the spectral distribution of the energy (color). It tells us, for example, that a star emits twice as much energy in the green as in the blue; but it does not tell us how much energy it emits in the green and hence in the blue. But though it is more satisfactory, the effective temperature, like the color temperature, is only an index of temperature. This is because, first, a star is not a blackbody, and, second, a stellar atmosphere, for the very reason that it is not in thermal equilibrium, has not one but many temperatures, which differ from point to point. It is the same with the earth's atmosphere, after all; in the same instant it can be freezing in New England and sweltering in Texas. Table 2 lists the effective temperatures of some of the brightest stars in the sky.

We have thus made another step forward. Apart from a few details we shall not go into, we are now able to determine stellar temperatures with pretty fair accuracy. And we have done it without using the classic thermometer and without even leaving the earth. Some people are apt to be suspicious of astronomical results because we cannot make "direct measurements." But, to make another digression, even here on earth not all measurements are done directly. Besides, what does "direct measurement" mean? In the case of temperature it usually means a measurement made with a mercury thermometer. This simply means applying a physical law that says that a body expands as its temperature increases. Placed in contact with a warmer body, the mercury in a thermometer tends to reach thermal equilibrium with it; consequently,

TABLE 2
Effective temperatures of the atmospheres of some bright stars

Star	Effective temperature (°K)
Sirius	10,200
Vega	10,700
Arcturus	4,300
Capella	5,000
Procyon	6,300
Altair	8,100
Aldebaran	3,500
Pollux	4,700

the mercury's temperature rises, and so it expands out of the thermometer's bulb and into its glass tube, which is appropriately graduated. And we read the temperature. But there are other ways of taking temperatures, based on other physical phenomena. Bodies would still have temperatures even if there were no mercury thermometers. If physics is the same on the earth and on the stars, that is, if physical laws are valid throughout the universe— pay attention now, because this is the foundation stone of our entire edifice—then all the thermal effects that enable us to derive the temperature are themselves the bases for thermometers. To "measure" stellar temperatures we have applied our knowledge of blackbody physics, and there is no question that the blackbody works as I described. Thus, once we have found the blackbody curve that fits our observations, we have indeed found the temperature (within the well-known limitations). I can think of only two reasons why people would still doubt our temperature measurements. Either they have not accepted the idea that physical laws are the same everywhere or they have not understood a thing. In the first case I do not know what to say. I would only point out to them that everything works so well on the basis of this assumption that in its place we would have to invoke an extraordinary number of coincidences. I would also remind them that we have concrete evidence (such as moon rocks and Mars's soil) that earth physics, once thought to be peculiar to the earth, applies to the planets as well. In the second case I would advise them to start again from the beginning or to find a book that explains things better.

To conclude, telescopic observations made with simple colored filters or a spectograph allow us to determine stellar temperatures.[38] Cooler stars are red and produce a spectrum dominated by the red (see Planck's curves), while hot stars are blue and produce a spectrum dominated by the blue. Plate 5 shows the spectra of a group of stars that are fairly similar to each other. This photograph was taken with the objective-prism technique, which consists in placing the prism in front of the telescope's lens. The light from each star in the field is scattered just before entering the lens, so that at the focal plane all the spectra can be seen at the same time. Plates 6 and 8 illustrate the different colors of stars. The photograph in plate 6 was taken with a long exposure time, and due to the earth's rotation, the stars (which in this case do not belong to a group) look like arcs rather than points.

Now we shall see how the temperature can help us to estimate the radius (the actual size) of those spheres of gas—the stars—that we can only see as faint specks of light. We know that the area delimited by the Planckian curve for a temperature T_e (effective temperature) represents the energy emitted in the unit of time by the unit of area (emissivity). We also know that by Stefan's law the emissivity is σT_e^4. Thus a star's total energy output is equal to its surface area times the energy emitted per unit area; that is,

$$L = 4\pi R^2 W = 4\pi R^2 \sigma T_e^4,$$

Where L is the luminosity, R is the radius, and W is the emissivity.

The sun's luminosity and radius, which are easy to calculate (relatively speaking), are usually taken as units of measure of stellar luminosities and radii. We know the sun's luminosity; if you recall, $L_\odot = 4\pi d^2 S$. We can also express the sun's luminosity in a different way by using the previous equation, that is,

$$L_\odot = 4\pi R_\odot^2 W_\odot,$$

where $R_\odot$ is the solar radius and $W_\odot$ the sun's emissivity. But then

$$4\pi R_\odot^2 W_\odot = 4\pi d^2 S,$$

whence we obtain the sun's emissivity:

$$W_\odot = S(d/R_\odot)^2.$$

It is important to note that we do not really need to know either d or $R_\odot$ but their ratio. The reciprocal of $d/R_\odot$ is $R_\odot/d$, which in turn, as shown in figure 42, is the angular radius of the sun α, that is, the angle (in radians) formed by two lines of sight, one to the center and the other to the edge of the sun. (Actually, it is the tangent to the angle, but it does not really matter since we are dealing with small angles.[39]) Once $W_\odot$ is known, we can easily find the sun's effective temperature T_e without looking at the Planckian curves; as we know from Stefan's law, $W_\odot = \sigma T_\odot^4$.

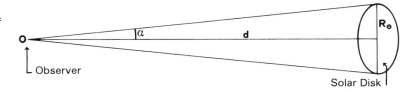

Figure 42
Diagram for the definition of angular radius (or diameter) of the sun (as well as the angular size of any object). α is approximately tan α, which equals $R_\odot/d$.

To continue with our determination of stellar radii, we know that for a star

$$L = 4\pi R^2 \sigma T_e^4$$

and for the sun

$$L_\odot = 4\pi R_{e\odot}^2 \sigma T_{e\odot}^4.$$

By dividing the first of these two equations by the second, we obtain

$$\frac{L}{L_\odot} = \left(\frac{R}{R_\odot}\right)^2 \left(\frac{T_e^-}{T_{e\odot}}\right)^4,$$

and hence

$$\frac{R}{R_\odot} = \left(\frac{T_{e\odot}}{T_e}\right)^2 \sqrt{\frac{L}{L_\odot}}.$$

Recalling that the luminosity is tied to the absolute magnitude by Pogson's relation (which is obviously valid for absolute magnitudes as well), it follows that by knowing a star's temperature and absolute magnitude, we can determine its radius.

Let us take as an example a red star with a surface temperature of 3,000°K and a luminosity 250 times that of the sun (equivalent to a difference of 6 magnitudes). If we write numbers in our last relation (without forgetting that our result will be approximate!), we obtain

$$\frac{R}{R_\odot} = \left(\frac{6{,}000}{3{,}000}\right)^2 \sqrt{250} \cong 4 \times 16 = 64$$

(the effective temperature of the sun is approximately 6,000°K). Thus the star's radius must be about 60 times that of the sun, which we know, and therefore it can be easily calculated.

To summarize, measurements made with colored filters or spectrographs enable us to get a fairly good idea of stellar temperatures, which, along with measurements of distance or absolute magnitudes, enable us to determine the sizes of stars. From radii and temperatures we can calculate luminosities (obtainable also from absolute magnitudes) by taking as reference the sun, whose luminosity and absolute magnitude are well-known.

·THE LINE SPECTRUM

Good. We are now done with the continuous spectrum. And what can the line spectrum tell us? Many things. I shall discuss some of them partly to underline the experimental basis of the astronomer's work and partly to prove to you that astronomers are second to none when it comes to extracting information from nature—in this case a ray of light.

In the first place, whether we invoke the reversing layer, spectral lines are produced by the atoms in a stellar atmosphere by absorbing and emitting radiation. Spectral lines, therefore, allow us to make a chemical analysis of the atmosphere; to put it simply, they tell us what a star is made of. To me (and to many other people as well) this is quite a significant piece of information. The first conclusion we reach is that all celestial objects consist (more or less) of the same elements that make up our small world (see table 3, which shows the chemical composition of three fairly different celestial objects). It is a clear proof of the physical and chemical unity of the universe.

TABLE 3
Relative abundance[a] of some elements in three celestial objects

Element	Sun (G2[b])	τ Scorpii (B0[b])	Diffuse nebulae (HII regions[c])
Hydrogen (H)	10^{12}	10^{12}	10^{12}
Helium (He)	$10^{10.9}$	10^{11}	10^{11}
Carbon (C)	$10^{8.6}$	$10^{8.2}$	$10^{8.7}$
Nitrogen (N)	$10^{7.9}$	$10^{8.3}$	$10^{7.6}$
Oxygen (O)	$10^{8.8}$	$10^{8.7}$	$10^{8.8}$
Neon (Ne)	$10^{7.9}$	$10^{8.6}$	$10^{7.9}$
Magnesium (Mg)	$10^{7.5}$	$10^{7.5}$	
Aluminum (Al)	$10^{6.7}$	$10^{6.2}$	
Silicon (Si)	$10^{7.6}$	$10^{7.6}$	
Iron (Fe)	$10^{7.3}$	$10^{7.3}$	

a. The number of atoms per cubic centimeter of the elements other than hydrogen is given in terms of a concentration of hydrogen atoms per cubic centimeter arbitrarily set at 10^{12}. Thus the table shows how much each element is more or less abundant than another. For example, in the sun helium is $10^{10.9}/10^{7.9} = 10^3$ times more abundant than neon; hydrogen is $10^{12}/10^{7.3} = 10^{4.7} = 50,000$ times more abundant than iron. (That is, for every 50,000 atoms of hydrogen there is 1 atom of iron.)

b. G2 and B0 are the spectral types of the sun and τ Scorpii, respectively. Spectral types will be discussed shortly.

c. HII regions will be discussed later.

• THE MOTIONS OF THE STARS: THE DOPPLER EFFECT

In addition to chemical composition, spectral lines reveal the speeds of stars and other celestial objects toward or away from the observer. To explain this point I shall use the classic example found in most astronomy books. At some time in your life you must have been overtaken by a car speeding by with its horn blaring; on many Italian highways it happens all the time. You must also have watched car races on television or seen a whistling train pass by. If you recall, the train's whistle becomes higher-pitched as the speeding train approaches and deeper as the train recedes. This effect is called the Doppler effect. The pitch of sound is determined by the frequency of the sound wave—a quantity inversely proportional to the wavelength: the shorter the wavelength, the higher the pitch; and conversely, the longer the wavelength, the lower the pitch. If the source is stationary, the number of sound waves that reach the ear is constant in time and the listener hears a sound of constant frequency (that is, wavelength); if it is do, re, or mi, it remains do, re, or mi. When the source is in rapid motion toward the listener, the individual sound waves tend to crowd up on each other because some of them have a shorter way to travel, and therefore a greater number fall upon the ear every second. The effect is the more conspicuous the closer the source comes and the faster it travels. For the listener it is as if the frequency increased. Thus if the approaching source emits a do, the listener hears a re or a mi. Conversely, when the source is moving away, the sound waves are stretched out, rather than compressed, and the pitch of the sound becomes lower.

An altogether similar effect operates in the case of light waves, and electromagnetic waves in general. If a light source is approaching, instead of the emitted radiation, the observer will see a radiation of shorter wavelength; and if the source is receding, a radiation of longer wavelength. In the first case the spectral line will be shifted from its normal position toward the violet, and in the second case toward the red. You will ask, How do we know that the spectrum has shifted from the position it would have had had the source been stationary? To begin with, I have to explain that the magnitude of the shift, known as Doppler shift, is not the same at all wavelengths. It is given by the equation

$$\Delta\lambda = \pm\lambda(v/c),$$

where $\Delta\lambda$ is the magnitude of the shift (that is, the amount that must be added to or subtracted from the wavelength λ of the emitted radiation to obtain the observed wavelength), v the velocity of the source, and c the speed of light. It follows that the intervals between the lines of a moving source are different from the corresponding intervals in the spectrum of a stationary source; furthermore, they vary, not at random, but proportionately to the wavelength. We could try to spot these regularities in a spectrum, but this would be a most laborious task. There is a much simpler and more accurate method that I shall illustrate with another example. Suppose we ask a pianist to play a simple tune while we stand next to the piano. If we tape the tune, we have recorded the sounds emitted by a stationary source. Next we move a long way from the pianist, whom we then ask to play the same tune on a piano mounted on a moving platform. As the piano races toward us, we again tape the tune. (We might need an amplifier, but let's not quibble!) When we compare the tapes, obviously it will not seem like the same tune. The pianist will probably cry foul having heard the same sounds both times. (It could not be otherwise; one always remains stationary relative to oneself.) But we have not heard the same sounds; the second time around the sounds we recorded were "shifted," and we can even calculate the speed of the platform by comparing the sounds on the two tapes.

Astronomers operate on the same principle, and although the shifts are small in the case of stars, the effects are often measurable. Suppose we place a laboratory light source before the slit of a spectrograph. This is our stationary source because clearly it does not move with respect to the observer (the spectrograph). If the source contains hydrogen, the spectrograph records the spectrum of hydrogen, and the hydrogen lines fall where their wavelengths dictate. Without touching any part of the spectrograph (or we change the observer!), we remove the laboratory source and let the light of a star fall on the slit next to where the light of the source fell. The hydrogen lines of the star will also show up on the focal plane (where spectra form) according to their wavelengths. If the star is stationary with respect to the observer, that is, if it does not move along the line of sight,[40] the lines of stellar hydrogen will line up exactly with the hydrogen lines of the laboratory source. If, instead, the star is moving along the line of sight, its hydrogen lines will be shifted to the red or the violet depending on whether it is moving away from or toward the observer.

The Doppler shift is not the same for the various lines; as the above equation tells us, it is larger at longer wavelengths. Consequently, to measure Doppler shifts it is more convenient to use lines corresponding to longer wavelengths. To get an idea of the amount of displacement let us measure a Doppler shift. Suppose we take a line at 6,000 Å in the red produced by a star with a radial velocity of 100 km/sec. By writing these numbers in our equation, we find that

$$\Delta\lambda = 6,000 \, \frac{100}{300,000} = 2 \text{ Å},$$

that is, a shift of 2 Å. If our spectrograph distributes the spectrum in such a way that the dispersion on the photographic plate is of 20 Å/mm (a very dispersive spectrograph, which can only be used for fairly bright stars),[41] 2 Å correspond to 0.1 mm. This means that given a radial velocity of 100 km/sec and using such a spectrograph, a spectral line of the wavelength considered shifts 0.1 mm with respect to the same line produced by a laboratory source. The difficulty of measuring such small displacements is often compounded by the fact that spectral lines are not very thin. And if we were to use a spectrograph with a dispersion of 200 Å/mm, the shift on the focal plane would be evn smaller; 2 Å in this case would correspond to 0.01 mm. Such a quantity is difficult to measure for purely technical reasons as well. Normal photographic emulsion has a grain (much like the grain of a newspaper picture) that does not allow us to distinguish, or *resolve*, details smaller than a minimum amount, which is different for different emulsions. The photographic emulsions generally used have a resolving power of 30–50 lines/mm, which means that we can only distinguish details up to 0.03–0.02 mm. Emulsions of higher resolving power exist, but to work they need a larger amount of light. In each observation, therefore, the best compromise must be found between different and often contrasting requirements.

To close, let us return to our equation once again. Obviously, once Δλ is known v can be easily calculated, since

$v = \Delta\lambda(c/\lambda).$

Thus if we can measure the Doppler shift, we can determine the radial velocity of the source. This means that we can obtain the radial velocity of all the stars whose spectra lend themselves to this type of measurement. As we shall see later, the use of the Doppler effect has made fundamental contributions to astronomical research.

·SPECTRAL CLASSIFICATION OF THE STARS

One would think that there must be as many different spectra as there are stars because no two stars can be exactly alike. This is probably true, but to verify it we would have to have stellar spectra as good as the sun's, which is not possible. We receive so much light from the sun that we can disperse and analyze its spectrum any way we like. But from the scant light we receive from the stars, we can only obtain fairly small-scale spectra that do not give us nearly as much information. In any case, we can try to put stellar spectra in some sort of order, which is just what Harvard astronomers did a few decades ago. It seems reasonable to start from the simplest spectra, which consist of a continuum with a few lines, and gradually introduce spectra that are more complex (have more lines) or show certain changes in the lines (for example, lines that intensify or fade away). The result of this work of spectral classification is the Harvard sequence, which groups stellar spectra into seven classes that comprise about 99% of all stars; the remaining 1% are stars whose peculiarities set them apart and will not be considered here. The seven spectral classes are labeled, in order, O, B, A, F, G, K, and M. This peculiar arrangement of letters has to do with the history of the sequence; at first, for example, there were reasons to believe that spectra of class A should come before spectra of class B, but later their order had to be inverted. English-speaking astronomers have found a way to remember the letter sequence with the sentence, "Oh Be A Fine Girl Kiss Me." (We Italian astronomers have adopted it, like so many foreign expressions.) To take into account more subtle differences, it has been necessary to divide each of the Harvard classes into ten subdivisions by affixing a number from 0 to 9 to each letter. Thus a star of spectral class A5 has a line spectrum whose pattern lies midway between those of class A0 and F0. In this classification the sun belongs to class G2.

Here is a brief description of the spectral classes. Class O displays lines of neutral hydrogen (HI), neutral and once ionized helium (HeI, HeII), and twice ionized oxygen, carbon, and nitrogen (OIII, CIII, and NIII). In class B the lines of neutral hydrogen and helium are stronger, while the lines of HII are absent. There are lines of OII, NII, and CII, as well as lines of twice ionized iron (FeIII) and magnesium (MgIII). In class A the hydrogen lines reach a peak of intensity at A2, then start getting weaker. The lines of neutral helium are weaker, while the lines of the metals are stronger.[42] In class F the hydrogen lines are noticeably weaker; they will rapidly fade away along the rest of the sequence. The lines of the metals grow even stronger. Class G spectra are dominated by the lines of the metals. The strongest lines are the H and K lines of once ionized calcium in the violet. These spectra display lines of the most common metals—iron (Fe), magnesium (Mg), titanium (Ti), calcium (Ca) and so on—in both the neutral and ionized states. Metallic lines of low excitation grow in number through classes K and M. In addition to atomic lines, we observe the broad bands of molecular compounds. Class M is dominated by the lines of titanium oxide.

Each spectral class from F to M can also be divided into two parallel branches, one denoted by the prefix g and the other by the prefix d. Thus we have types gF, gG, gK, and gM, characterized by thin lines, and types dF, dG, dK, and dM, characterized by broad lines. The lines are the same since the spectral types are the same.

Taking a look at the continuum, we find that it is very intense in the violet in spectral classes O and B; then the emission peak shifts gradually toward the red through classes A, F, G, K, and M.

A very important aspect of the spectral classification is that it is also a temperature sequence. O and B stars are intrinsically blue, A and F stars white, G stars yellow, K stars orange, and M stars red. As we know, hot stars have maximum emission in the blue and cooler stars in the red; thus surface temperatures decrease gradually along the sequence from O stars to M stars.

Measurements of effective temperatures show, approximately, $40 \times 10^3 \div 50 \times 10^3 \,^\circ K$ for type O stars, $20 \times 10^3 \div 30 \times 10^3 \,^\circ K$ for type B stars, $10 \times 10^3 \,^\circ K$ for type A stars, $5,500 \,^\circ K$ for type G stars, $4,500 \,^\circ K$ for type K stars, and $2,800 \,^\circ K$ for type M stars.

Looking back at our spectral classification, we find that in the visible part of the spectrum (between 4,000 Å and 7,000 Å, which is the wavelength interval considered by the Harvard astronomers) the outstanding features in class O spectra are the helium lines, in class A spectra the hydrogen lines, and in class M spectra the lines produced by the metals and molecular compounds. We might be tempted to conclude that stars are different in composition—to make an extreme case, that O stars are rich in helium and devoid of metals, while M stars are rich in metals and devoid of helium. But it would be the wrong conclusion. We are apt to make the same sort of mistake in everyday life. Given a set of data, we consider only those data in the set that seem most important and draw our conclusions accordingly. In retrospect it turns out to be a mistake, and some of these mistakes may even condition our lives for a long time, if not forever. Eventually we realize that what seemed important was not and that proper consideration should have been given to all the data; had we done so, we would have saved ourselves a lot of pain. In the case of stellar spectra the mistake is not so tragic, but it is good to remember that one should not be too hasty. On the other hand, one should not be cautious to the point of being unable to make decisions. Thinking out a problem does not mean agonizing over the choices, but rather learning to give proper weight to all the factors involved and, if possible, to simplify the problem.

In this context it may be useful to underline one of the fundamental aspects of scientific research. Whether it is good or bad, whether it benefits mankind or results in great evils, as many believe, scientific investigation since Galileo's time has been governed by some strict rules. One of those that it must satisfy is the rule of separation of the variables: Every natural phenomenon is the result of a sum, an overlapping, of interactions of various effects having various causes. It is almost impossible to explain a phenomenon as a whole, the way we observe it. One example of this is the inability of the people (not at all stupid) who lived before 1600 to give adequate explanations of natural phenomena; another example is the poor, arbitrary interpretations given by those of our contemporaries who are unable or unwilling to understand and apply the scientific method to the investigation of nature. Separation of the variables is an essential part of the scientific method. It consists in identifying the variables at play, that is, the "simple" causes, the "simple" phenomena whose simultaneous occurrence produces a complex phenomenon. Identification of the simple causes often leads quickly to their separation into simple fundamental causes and simple accessory causes; the latter are also important in that they help to produce a phenomenon that cannot be exactly repeated in its entirety, but they are not what produced the phenomenon in the first place. Let us use a falling stone as an example. If on a windy day we let a stone fall from the same point several times, it is very unlikely that the stone will *always* fall in the same way and *always* hit the same spot. But it will surely fall even on a windless day. This means that in the falling-stone phenomenon the wind is an accessory cause, a nonessential factor that we ought to disregard; we would take it into account only if we wished to solve a specific case— the fall of a given stone in a given wind. Similarly, the trajectory of a cannon ball must first be determined for "ideal conditions" (no wind, perfectly shaped barrel, and so on). It is only when we come to the actual firing that we must take into consideration the various factors, such as wind strength, that can alter the trajectory. In the case of the falling stone, after all accessory causes have been removed, the only thing left that is *essential* to the understanding of the phenomenon is the force of gravity. All this may seem trivial and obvious now, but it took thousands of years to get there. Once this concept—and a few others—were understood, the scientific method was born and science knew an explosive growth.

Coming back to our case, the fact that the various spectral classes display lines of different elements does not entitle us to conclude that such variations are due to changes in chemical composition. Basically, it just says that O stars are hotter than M stars, something we already knew. The temperature of O stars is so high that the metal atoms are all ionized—the bond between electrons and nucleus is very weak in these atoms, and it does not take much to free one or two electrons—and the spectrum of an ionized atom is completely different from that of a neutral atom. If the transitions of the ionized atom fall in the ultraviolet, no lines will show up in the visible region of the spectrum. The helium atom, instead, is very difficult to ionize. This is why the spectra of hot stars show helium lines but no metallic lines. Conversely, at low temperatures the helium atom cannot be excited and is essentially at its lowest energy level. Hence the helium lines are absent from the spectra of cooler stars. But the lines of neutral metals are there, for reasons that are now obvious, and so are the lines of molecular compounds, since the forces that bind atoms together in a molecule, though weak, are strong enough to withstand the forces operating in a low-temperature medium.

Thus the spectral variations we observe are due mainly to changes in temperature. The relative abundance of the elements can and does vary from one star to another, but it is a secondary factor to be taken into account only when we are interested in particular effects or subtler classifications.

It now remains to explain why, starting with F stars, spectral classes are subdivided into g and d types. If you recall, a rarefied gas produces a thin-line spectrum and a dense gas a broad-line spectrum. The subdivision takes into account the fact that starting with class F there are both low-density and high-density stars.

To summarize, the appearance of a stellar spectrum is determined chiefly by temperature; in much smaller measure, it is also determined by density and chemical composition.

• A CORNERSTONE OF MODERN ASTRONOMY: THE HERTZSPRUNG-RUSSELL DIAGRAM

Let us now follow in the footsteps of E. Hertzsprung and H. N. Russell, the two astronomers who in 1911 and 1913, respectively, made a discovery that gave us a new insight into the world of stars. It did not happen so long ago (at that time my father was already in his teens); yet in the few years since, the window opened on the universe by Hertzsprung and Russell has enabled us to look at it in a new way and to follow a broad avenue of research, not yet fully explored, that has yielded the key to stellar evolution.

Let us start from the beginning, and the importance of their discovery—truly a cornerstone of modern astronomy—will emerge from the story. Using stars of known distances, Hertzsprung and Russell constructed a diagram in which spectral types[43] were plotted against absolute magnitiudes, as shown in figure 43. Each star is thus represented by a point whose horizontal coordinate is an indication of the temperature and whose vertical coordinate is a measure of the luminosity, or total energy flux. Spectral classes are plotted from left (O stars) to right (M stars); since the spectral sequence is a temperature sequence, the temperature decreases from left to right. Absolute magnitudes are plotted in decreasing order from top to bottom.

Let me emphasize that we are talking of *absolute* magnitudes; therefore, the stars at the top are intrinsically brighter than those at the bottom. Recall that 5 classes of magnitude correspond to a 100-fold increase (up the scale) in the energy flux; thus a star of absolute magnitude 0 is intrinsically 10,000 times brighter than a star of absolute magnitude +10.

Figure 43
The Hertzsprung-Russell diagram. Three interchangeable scales are given on the abscissa for comparison—spectral type, temperature, and color index.

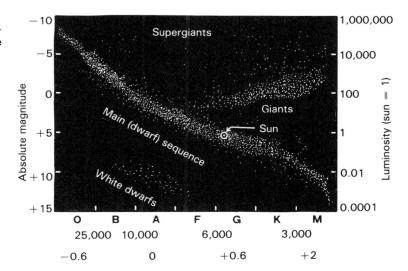

Even without developing complicated theories, one might suspect that there must be a correlation between temperature and luminosity. And indeed there is, as figure 43 shows. You will notice that the points are not scattered at random over the diagram, but tend to be concentrated in well-defined zones. The correlation cannot be a simple one, however, considering that the points fall in various groups rather than in a single line or strip, and it needs to be investigated.

I cannot give you a detailed history of these studies. It is not at all simple, and its detailed treatment would give a completely different slant to the book. But you must have the idea by now that astronomical research proceeds by trial and error, by making models and perfecting them as new experimental and theoretical tools become available. And so it was for the Hertzsprung-Russell diagram. The results of this research will be briefly summarized in part III, on stellar evolution. As always, however, you must bear in mind that current views should not be regarded as definitive, but rather—more modestly—as the best we can give for now. Most of the things that are taught today as a matter of course, as if they had always been known, were not taught, or known, when I was a university student.

We should pause a moment, now, and examine the Hertzsprung-Russell diagram, or the H-R diagram as it is called for the sake of brevity and the love of acronyms. Keep in mind that the diagram is constructed with stars whose distances can be measured by the trigonometric method; hence it only refers to stars in the sun's neighborhood. As a general rule, you must always remember the observational bias, and in this particular case you should not be surprised if the diagram changes for different groups of stars. One never knows. The job of researcher does not promise you great discoveries, which are the privilege of a few; but one thing it teaches you is to accept the unexpected and to confront things with an open mind.

Unfortunately, not everyone learns this lesson. The scientific profession, like any other, has its share of dumb, petty, arrogant, dishonest, incompetent, and foolish people. Do not be too impressed with titles and academic credentials; sometimes—not always, of course, or there would not be any science to speak of— positions, titles, and honors are awarded to people whose greatest accomplishment is to belong to the right circle and who are willing to pay their entry dues. (The same goes for literary prizes and other laurels.) It is a delusion fed mainly by the interested parties that a scientist is ipso facto a better person. In reality, the scientific community, like any other, is composed mostly of mediocrities— useful, indeed, indispensable, but still mediocrities. Not that you could tell the way they strut about in universities and research institutions, in lecture halls and on television screens.

We had better return to the H-R diagram. The first point is that the diagram includes stars of all spectral classes. The second point, the great majority of stars fall in a belt that runs diagonally across the diagram from top left to bottom right. This belt contains all spectral types, from blue, hot, very luminous stars at the top to red, cool, very faint stars at the bottom. This continuous band of stars is called the *main sequence* (also the dwarf sequence). The sun, a dwarf star[44] of spectral type G2 and absolute magnitude +4.8, is on the main sequence. The rest of the stars are concentrated in different zones. Above the main sequence are two groups of exceptionally luminous stars: the giants, which extend from spectral class F to spectral class M and have absolute magnitudes from 0 to ±1; and the supergiants, which are sparsely scattered across the top, include all spectral classes, and have magnitudes from −3 to −7. The latter are so bright that they can be seen even at great distances (hence many of the stars that are visible with the naked eye are supergiants; see figure 84). At the opposite end of the diagram, far below the main sequence, lie the so-called white dwarfs—hot but very faint stars of spectral classes B, A, and F. Other groups have been identified—subdwarfs, subgiants— but we do not need to concern ourselves with the finer details. The main point of the H-R diagram is that there is a correlation between absolute magnitude (luminosity) and spectrum (temperature) and that this correlation is not a simple one. Since the correlation between the two quantities is not a line but a strip of a certain width, there may be a further complication, or, alternatively, the measurements made for the individual stars may be in error; the meaning of the diagram clearly depends on our ability to settle this point. Finally, it should be noted that the groups mentioned are the zones of greatest concentration of the stars. But there are points scattered about almost the entire diagram. Are these points due to the grossly wrong measurements or are they as legitimate as the points that fall in the main groups? And if they are "right," what do they mean? To find the answers to these questions, you will have to wait until the next chapter. (A little suspense won't hurt!) Meanwhile, let us see what we can do with the H-R diagram.

Suppose we have two stars of spectral class K, one a dwarf and the other a supergiant. Suppose further that their absolute magnitudes are $+6$ and -4, respectively. The supergiant is thus 10,000 times brighter than the dwarf. Since the two stars have the same temperature (they belong to the same spectral class), the energy radiated per unit of surface area must be the same for both stars. It follows that the supergiant is not only brighter but actually larger than the dwarf. How much larger can be easily calculated. If the supergiant is 10,000 times brighter and if the flux per unit area is the same, then the radiating area of the supergiant must be 10,000 times larger than that of the dwarf. Stars being spheres to all intents and purposes, the ratio of the two stars' areas is equal to the square of the ratio of their radii. Thus the ratio of the radii is equal to the square root of the ratio of the areas and, in our case, to the square root of the luminosity ratio. It follows that the diameter of our supergiant is 100 times that of the dwarf.

The main star in the binary system α Herculis is a red supergiant whose luminosity is about 15,000 times that of the sun. But its diameter is more than 120 times that of the sun because it belongs to spectral class M5; and since it has a lower surface temperature (2,700°K), its emissivity is lower. The diameter of the supergiant is actually about 600 times that of the sun. Such a star could easily engulf a good portion of the solar system known to the ancients.

Without quite understanding the implications of the H-R diagram, we have been able to use it to get an idea of the sizes of stars. And we have done it without making direct measurements.

Therefore the knowledge that direct measurements in astronomy are very difficult—and generally impossible—need not dampen our hopes.

Indeed, one such discovery follows immediately from what has been noted. To construct the H-R diagram we have only used stars of known distances because we need distance to calculate absolute magnitude. Suppose now we have a star that is further away than 50–100 parsecs; the star's parallax cannot be measured by the trigonometric method, and therefore its distance cannot be calculated. But suppose we can obtain the star's spectrum and establish its spectral class. Starting from the spectral class on the horizontal axis, we draw a line in the upward direction and parallel to the vertical axis until we reach one of the zones in the diagram. From this point we draw a horizontal line across the diagram to the left until it intersects the vertical axis. The point of intersection tells us the absolute magnitude of the star even though we do not know its distance. But once we know the star's absolute magnitude and its apparent magnitude (which can be determined from the simple fact that the star is visible), we can calculate its parallax and hence its distance.[45] In other words, given spectral classes and apparent magnitudes, we can use the H-R diagram to calculate stellar distances—at least for all the stars whose spectra can be obtained. Parallaxes determined by this method are called spectroscopic parallaxes. And that is not all. What happens when a star is too faint for us to get its spectrum? By photographing it with different filters we can obtain a color index. The color index, too, is a temperature index, and we can certainly convert the spectral-class scale to a color-index scale. Thus if we can just photograph a star, we can calculate its distance. Isn't it fantastic? It's so simple, it's almost incredible. Of course, our figures cannot be too precise, since the diagram has strips rather than lines, but there is a whale of a difference between saying a vague "very far" and giving a number that, although imprecise, differs from the actual distance by some definite amount. It is all very simple now, but we must not forget that it could not have been done without the H-R diagram. We owe a great debt of gratitude to the host of patient souls who spent years measuring parallaxes, taking spectra by night, and classifying them by day. Without their combined efforts we would still be wondering how large the universe is. The very fact that we have begun to understand and explore the universe is due to the powerful tool they gave to astronomical research. All kinds of marvelous things followed, but we ought to remember the pioneers; they did the heaviest work.

Finally, I should clarify a point that I am sure my observant reader has not missed. We start from the spectral class and draw a line upward until we reach one of the zones in the diagram. Fine. But which zone? Where do we stop? Given a K star, we can stop at the main sequence, around absolute magnitude $+6$, or further up at the giants, around absolute magnitude $+1$, or even further up at the supergiants, around absolute magnitude -5. Obviously the value of the parallax varies considerably depending on where we stop. Quite right. The choice of a zone is determined by spectroscopic criteria that will be explained later, after we find out about stellar masses (of which we know nothing right now). Such criteria have to do with the widths of the spectral lines, which are broad for dwarfs, thin for giants, and very thin for supergiants.

· A BRIEF HOMILY

Aside from the importance of the results, it is worth remarking on another aspect of our voyage of exploration. By the power of reason and with the knowledge garnered from many different, fields, we have been able to understand things that might have seemed forever incomprehensible. You will have noticed that in astronomy we make use of everything—mathematics, physics in all its branches, and chemistry; they all play a part. It is a well-recognized fact today that astronomy is perhaps the most inter-disciplinary of all the sciences.

The fact that we can understand, that is, reason things out, make articulate statements, describe phenomena in ways that are intellectually satisfying, build models, make predictions that can be verified, identify the fundamental elements that bring order to near-chaos—this is what is truly rewarding about astronomy, or any other science. This is what separates human beings from the creatures who are at the mercy of incomprehensible natural phenomena. It does not matter that we can never be sure of saying the final word on any subject; in fact, we can be sure that the final word never will be said. What is important is the mental habit, a basic faith in our ability to think for ourselves. And reason is the only thing that can dispel prejudices and all other forms of bigotry. If we can build a reasoned view of the world, we can throw out every (let's say *almost* every) form of superstition, be it religion, astrology, palm reading, or white, black, or blue magic— in other words, all those things that are founded on gratuitous statements, never proved and not provable. All these sad, murky, frustrating things are apt to disappear upon experiencing the limpid joy of intellectual, conscious discovery.

•STELLAR MASSES

There is only one way to determine directly the mass of a star: apply a known force to it and measure the acceleration it undergoes. And there is only one kind of force that can affect the uniform (that is, constant-velocity) rectilinear (that is, straight-line) motion of a star in space: the force of gravity. But for a body to be subject to the force of gravity, there must be at least another object close by. Two bodies attract each other with the same force, which is given by Newton's law,

$$F = G \frac{m_1 m_2}{r^2},$$

where G is the constant of gravitation, m_1 and m_2 are the masses of the two bodies, and r is the distance between them. (Newton's law is for point masses. When two masses are spherical and homogeneous, r is the distance between their centers; if they are irregularly shaped, they must be far enough apart that they may be considered points, in which case r is the distance between them.)

The earth attracts a stone and the stone attracts the earth. They attract each other with the same force and move toward each other. Of course, everybody can see that it is the stone that falls to earth and not the other way around, but they do actually move toward each other. The point is, the force F applied to a mass m causes the latter to undergo an acceleration (change in velocity) equal to F divided by m. Since m is very large in the case of the earth and very small in the case of the stone, the acceleration that the stone impresses on the earth is as many times smaller as the mass of the earth is greater.

The force of attraction depends not only on the product of the two masses but also on the distance between them. More precisely, it decreases as the square of the distance. At 10 times the distance, the attractive force is 100 times less; at 1,000 times the distance it is 1 million times less. As for the constant G, it has a very small value ($G = 6.670 \times 10^{-8}$ dyne cm^2/g^2; 1 dyne is the force required to accelerate 1 g at the rate of 1 cm/sec^2).

All in all, the masses must be fairly large for the force of gravity to be appreciable. If separated by a great distance, however, even two large masses attract each other with a force that is only a number on a piece of paper. (Theoretically, two masses should affect each other at any distance. The force becomes zero only when r is infinite.) But in reality, beyond a certain distance the force of gravity is negligible. In the absence of the force of gravity (as well as any other force), a mass's motion undergoes no change—a mass at rest remains at rest; a mass in uniform rectilinear motion remains in uniform rectilinear motion. (There is no reason for a mass to start moving in the first case or to change its motion—to accelerate, for example, or to veer off in one direction or another—in the second.)

Stellar distances (as opposed to planetary distances) are so great that no star by itself can affect the motion of other stars (the reason why is clear from the above discussion of the force of gravity). Thus if all the stars were separated by such great distances, there would be no way of measuring stellar masses. But luckily there are stars close enough to each other to be gravitationally bound.

· VISUAL BINARIES

In 1650, shortly after Galileo's death, Giovanni Battista Riccioli observed through his telescope that Mizar, a star in Ursa Major, seemed to consist of two stars. Obviously, two stars that are close by in the telescopic field are not necessarily close to each other in space; if they are in the same line of sight they will appear close together even though millions of miles apart. But Mizar, as it was later ascertained, is actually a system of two stars in close proximity, physically tied to each other, and was the first double star (or binary system) observed by man. Many others were subsequently discovered. From 1782 to 1821 William Herschel cataloged more than 800 "double" stars, most of which turned out to be true doubles rather than "optical" doubles, as could have been expected. If pairs are observed in large numbers, it is reasonable to expect that many of them will really be double systems. Assuming that the stars are distributed at random, we can calculate the probability that two of them should appear so close together that only the telescope can tell them apart; the probability is very small.

In 1804 W. Herschel noticed that the fainter star he had observed near Castor (α Geminorum) had changed its position in relation to Castor. He concluded that Castor had a companion revolving about it. This was the first observation that gave universal validity to Newton's law of gravitation. (If you recall, Newton's *Principia* appeared in 1687.) From that time on the law of gravitation was valid, not only on the earth and in its environs, but throughout the sky, where the "fixed stars" once had been. This is why 1804 is one of the important dates in the history of astronomy.

Herschel's son John carried on his father's work. During his lifetime he cataloged more than 10,000 multiple-star systems.

The first binary orbit was determined by F. Savary in 1827 for the star ζ Ursae Majoris. The orbit of one star about the other is an ellipse with a period of 60 years.

Today we know that double stars are very numerous—indeed, many more than one would have imagined; almost half of the stars are binary systems. I shall leave it to your imagination to picture how the sky would appear to the hypothetical inhabitant of a hypothetical planet in a double, triple, or quadruple system in which the suns are stars of different colors. The importance of the binary systems is that they enable us to take the first step in determining stellar masses. Once we have measured distances, temperatures, dimensions, and masses, we shall be well on our way to understanding the world we live in.

Take a step backward for a moment. Let us see how we can determine the mass of the planet Mars. By means of an accurate clock and a telescope equipped with a micrometer (or camera), we can measure Deimos's orbital period P and its average distance a from the center of Mars. We find that $P = 1.262$ sidereal days and $a = 23,450$ km. Next we turn to the earth-moon system and we measure the same quantities for the moon's orbit around the earth; in this case $P = 27.3$ sidereal days and $a = 385,000$ km. Kepler's third law (in its exact form; compare note 2) states that

$$a^3 = \frac{G}{4\pi^2}(m_1 + m_2)P^2,$$

where m_1 and m_2 are the masses of the two bodies gravitationally bound to each other (the earth and the moon, Mars and Deimos, the sun and the earth, Jupiter and one of its satellites, and so forth). Writing the third law first for the Mars-Deimos system and then for the earth-moon system, we obtain

$$a_D^3 = \frac{G}{4\pi^2}(m_M + m_D)P_D^2$$

and

$$a_L^3 = \frac{G}{4\pi^2}(m_E + m_L)P_L^2,$$

where M stands for Mars, D for Deimos, E for the earth, and L for the moon (Luna). Thus

$$\frac{a_D^3}{a_L^3} = \frac{m_M + m_D}{m_E + m_L}\frac{P_D^2}{P_L^2}.$$

Writing in the numbers we found earlier, we obtain

$$\left(\frac{23,450}{385,000}\right)^3 = \frac{m_M + m_D}{m_E + m_L}\left(\frac{1.262}{27.3}\right)^2;$$

hence

$$(0.061)^3 = \frac{m_M + m_D}{m_E + m_L}(0.046)^2;$$

that is,

$$\frac{m_M + m_D}{m_E + m_L} = \frac{(0.061)^3}{(0.046)^2} = \frac{0.00023}{0.00212} = 0.108,$$

which can also be written

$$m_M + m_D = 0.108(m_E + m_L).$$

Disregarding m_D and m_L, which are negligible compared to the masses of Mars and the earth, respectively, we find

$$m_M \cong 0.108 m_E.$$

In words, the mass of Mars is approximately one-tenth the earth's. Thus if we know the earth's mass, we can calculate Mars's. Do you realize what has happened? By making a number of geometric and temporal measurements (positions, parallaxes, and orbital periods) and using Kepler's third law, we have determined Mars's mass in terms of the earth's mass. It seems incredible, but it is really very simple because Kepler's third law gives the relation between the sizes of the orbits, the orbital periods, and the masses of the two bodies.

I have to add that the mass of the earth can be calculated by a very delicate, but conceptually simple, method that requires the use of an extremely sensitive scale. It was done by P. von Jolly in 1881, but I shall not describe his experiment. I hope some of my readers will feel motivated to look it up in another book. Needless to say, the same way we determine the mass of Mars we can calculate the masses of all the other planets.

Let us return to double stars. It must be clear by now that their masses can be calculated by an analogous procedure. Here is how one goes about it. Observe figure 44 (which obviously is not to scale—or I would have not been able to draw it—and is also simplified). S is the Sun, E the earth, and A the earth-sun distance; M is the main star in the binary system, C its companion, and a the distance between them; d is the distance of the binary system M-C from the sun (or the earth) and is much (very much) greater than either a or A. The larger of the two ellipses represents the earth's orbit around the sun, and the smaller the orbit of C around M. The observer at S (or at E, which is the same thing because of the distances involved) measures the angle a'' formed by the lines SM and SC. This angle, so small that it is measured in seconds of arc, is called the *angular radius* (or angular semimajor axis) of the binary's orbit. As we know, the angle p'' subtended by the semimajor axis of the earth's orbit is the parallax of the double star.

Practically speaking, the ratio a/d gives the value in radians of the angular radius, and since there are 206,265 seconds of arc in a radian ($1° = 3,600''$ and 1 radian $= 57.29+°$; see note 11),

$$a'' = (a/d)206,265.$$

Analogously, the value of the parallax in seconds of arc is

$$p'' = (A/d)206,265;$$

hence

$$\frac{a''}{p''} = \frac{a}{A}.$$

Thus if we measure a'' and p'' separately, which we can do (note that we do not need to measure a in kilometers), and divide them, we can find a/A. And if we set $A = 1$, that is, if we take the earth-sun distance as unit of measure (recall that this distance is in fact called an astronomical unit), the ratio a''/p'' gives us the value of a in astronomical units. Naturally, once we measure the earth-sun distance in kilometers we can also calculate the semimajor axis of the binary's orbit in kilometers.

Figure 44
Diagram, not to scale, showing the orbit of the earth (E) around the sun (S) and the orbit of the secondary star (C) around the primary (M) in a binary system. *A* and *a* are, respectively, the semimajor axes of the two orbits; *d* is the distance of the binary system from the sun; *a"* is the angular radius of the binary orbit; and *p"* is the parallax of the binary system.

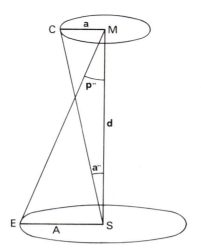

Suppose now that we have observed the double star long enough to make an accurate measurement of the orbital period P of the companion star about the main star. We shall take one year, T, as the unit of time and the mass of the sun, m_S, as the unit of mass. Nothing prevents us from doing this, and it simplifies our calculations. As I said earlier, we can always convert the distances to kilometers, the times to seconds (by knowing how many seconds there are in one year), and the masses to kilograms (by knowing the mass of the sun in kilograms).

Now we write Kepler's third law for the double star and for the sun-earth system

$$a^3 = \frac{G}{4\pi^2} (m_M + m_C)P^2$$

and

$$A^3 = \frac{G}{4\pi^2} (m_S + m_E)T^2.$$

Dividing the two equalities as we did before, we obtain

$$\frac{a^3}{A^3} = \frac{m_M + m_C}{m_S + m_E} \frac{P^2}{T^2}.$$

If $A = 1$, $T = 1, m_S = 1$, and $m_E = 0$ (we disregard the mass of the earth in relation to that of the sun), we can write

$$m_M + m_C = \frac{a^3}{P^3},$$

or

$$m_{\mathrm{M}} + m_{\mathrm{C}} = \frac{a''^3}{p''^3 P^2}.$$

Thus knowing the angular radius a'', the parallax p'', and the period P, we can calculate the *total mass* of the double star. We still have not determined the individual masses, but this cannot always be done; in this case there are ways of estimating the masses. But many times it can be done. It can be done, for example, when we can measure the absolute orbits of the two stars about their common center of mass, rather than the relative orbit of the companion about the main star. (It is about the common center of mass that the components in actuality revolve.) By observing the motion of the components in relation to background stars, we can find the semimajor axes a_{M} and a_{C} of the actual orbits about the center of mass. Since a_{M} and a_{C} are inversely proportional to the masses, we can write

$$a_{\mathrm{M}} : a_{\mathrm{C}} = m_{\mathrm{C}} : m_{\mathrm{M}}, \quad \text{with} \quad a_{\mathrm{M}} + a_{\mathrm{C}} = a.$$

Together with the preceding equation, this equality of ratios is enough to solve the problem.

In multiple systems we find all kinds of stars—large and small, hot and cool; as a result, we can determine masses for most types of stars. In the end we have a range of masses, from the heaviest stars to the lightest. In other words, we can find the upper and lower limits to the possible values of stellar masses.

At this point (I wonder why!) I feel you might need a pat on the back. Please, do not get discouraged. Do not say, "Heavens, what stuff. This is all too difficult." What I am giving you, though simplified, are the basic elements you need in order to understand the world of stars and the way our knowledge of them is acquired. And if you bear with me, you will find that none of these things are really beyond your comprehension, as you might have thought. One step at a time, we are composing a mosaic in which every piece has a precise meaning, a precise function in the overall composition. Some of the facts, observations, techniques, and phenomena may seem unrelated to the others, but when all the pieces finally fall into place, the result will be truly satisfying for its innate coherence. Personally, I find it very rewarding even from an esthetic point of view. I hope it is the same for you. I hope for your sake that you are capable of enjoying intellectual pursuits. It is such a pleasant feeling, satisfying yet stimulating, that if you cannot experience it, you will be missing one of the best things in life. You might argue that so many of the best things in life are denied to you—sailing, owning a castle, the ability to sing or play the piano, skindiving, running the 100 meters in less than 10 seconds, who knows what —that one more won't make any difference. I am sorry if you feel this way. This attitude implies a fatalistic view of life, in which humans are passive objects rather than active participants who create themselves, their histories, and their values. Sailboats may sink; castles can be confiscated; races are lost. But the world of ideas and intellectual pursuits, making your own choices, the teamwork that leaves behind an enduring legacy—these are things that nobody in the world can ever take away from you. This is what the original sin was all about. Adam and Eve had every material thing they needed. But it was not enough. As *human beings* they wanted knowledge. And we should not forget that it still is our most valuable possession.

Back to the stars. Double systems that are recognized visually, that is, with the telescope, are known as visual binaries. It is evident that either visual binaries are very close to us or they are composed of widely separated stars (otherwise we would not be able to see the components separately).

•PHOTOMETRIC DOUBLES, OR ECLIPSING BINARIES

Seeing its individual components is not the only way of identifying a binary system. Direct observation is not always necessary in order to substantiate a fact. Far be it from me to suggest that statements should be accepted "on faith" without any kind of direct or indirect proof; but it is important to realize that we cannot be limited to direct proofs. In astronomy, actually, we have to rely a great deal on indirect proofs—out of necessity, not by choice.

If the components of a binary system are so close together that not even the most powerful telescope can tell them apart, we see them as one star. Nevertheless, there are two of them. Whether it is one or two certainly does not depend on our being able to distinguish them. Suppose that the orbit of one of them is such that the star passes in front of the other in the line of sight and, after a half-revolution, behind it. We do not see it happen because we cannot see the stars separately. But we do see their light. And we receive less light when one star is in front of the other. In other words, the passage of one star in front of the other produces an eclipse. This is why such double stars are known as eclipsing binaries. If the stars are of equal size and brightness, at each eclipse the amount of light received on earth will be cut in half. Usually, however, the stars have different emissivities; that is, one of the stars radiates less per unit area than the other. (Let us call them the secondary and the primary, respectively.) When the secondary eclipses the primary, the loss of light will evidently be greater than when the secondary is occulted by the primary.

The first eclipsing binary to be observed was Algol, whose brightness fades by two-thirds approximately every 2.87 days and returns to normal in a few hours. It was first noticed in 1669— much too early to understand the reason for the variations in brightness. In 1783 J. Goodricke gave two possible explanations for Algol's periodic dimming; he suggested that it could be due to either a large cold area on the star (a spot) rotating along with the star or an invisible, dark companion periodically occulting the main star. It took another century for H. Vogel to identify the correct reason. With modern instruments the flux of radiation from the stars can be measured with great accuracy, and a large number of eclipsing binaries have been discovered. If the energy flux is recorded over a period of time, a light curve can be obtained by plotting the observed brightness against the time. Although different for different eclipsing binaries, the light curves are essentially similar in that they all show a periodic dimming of the total light. Two such curves are schematically illustrated in figure 45. The meaning of the diagram is explained in the caption and I do not think additional comments are needed. Nor would it serve any purpose to explain in technical details how these light curves are analyzed. Suffice it to say that from an accurately determined light curve (with suitable hypotheses about the shapes of the stars and our knowledge of stellar atmospheres and gravitational fields), we can obtain the geometric parameters that define the orbit of the secondary about the primary.

·SPECTROSCOPIC BINARIES

Consider now the case of a binary system in which the components are very close together but the orbital plane is so tilted that the stars never occult each other. There are no eclipses, hence no periodic variations in brightness for us to observe. This much is clear. Our ingenuity is not yet exhausted, however. For simplicity's

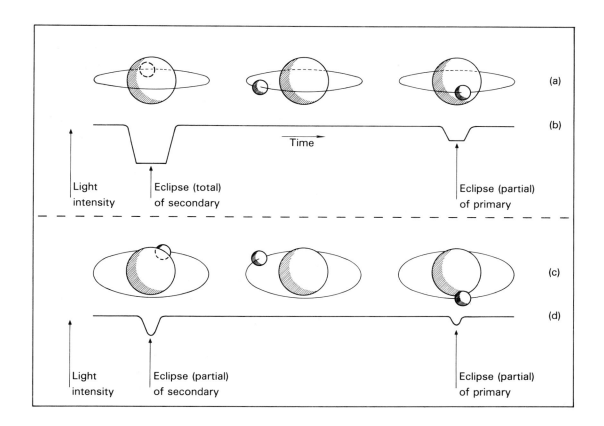

Light intensity

Eclipse (total) of secondary

Time

Eclipse (partial) of primary

Light intensity

Eclipse (partial) of secondary

Eclipse (partial) of primary

Figure 45
Parts (a) and (c) show the relative positions of the stars in two eclipsing binary systems. In (a) the orbit is slightly inclined on the line of sight, while in (c) it is much more tilted. If the orbit was even more tilted, there would be no eclipses. Parts (b) and (d) show schematically the corresponding light curves, that is, the fluctuations in the intensity of the light signal received by the observer. In both cases the light is dimmed less when the larger star is eclipsed. This means that the larger star is less bright, in the sense that each square centimeter of its sur-face emits less (because its surface temperature is lower). But it does not nec-essarily mean that the star as a whole is less luminous; it all depends on the size of the star (many square centi-meters, each not very bright, can collectively give much more light than a few very bright square centime-ters). Needless to say, there are binary systems in which the primary is the brighter star.

sake, let us say that the secondary revolves about the primary (rather than that both revolve about their common center of mass). Unless it revolves in a plane perpendicular to the line of sight—clearly an exceptional case—the secondary must periodically approach and recede from the observer. Due to the star's radial velocity, its spectral lines will show a Doppler effect. As you may recall, the Doppler effect is the more appreciable the higher the (radial!) velocity. In our case, the effect cannot be negligible. Since the stars are very close together (or we would see them separately) and since they do not fall onto each other, their orbital velocities must be very high. A double star that is recognized from the spectrum is known as a spectroscopic binary. Thus to observe these binaries we need a spectrum. This limits the number of stars that can be studied by this technique because very faint stars do not produce spectra that lend themselves to accurate measurements. But their number is not that small, and when all the results are combined, we have sufficient data about double stars to carry out statistical studies.

What exactly do we observe in the spectrum? Figure 46 illustrates schematically four typical positions of the components of a spectroscopic binary. In configuration (a) star A is approaching the observer, while star B is receding. The stars are assumed to have the same temperature and to produce the same spectrum. Due to the Doppler effect, all spectral lines of A will be shifted to the violet (with respect to the lines of a laboratory source) and all spectral lines of B will be shifted to the red. Of course, we cannot obtain separate spectra because for us the two sources are a single source; thus the spectrum we photograph results from the superposition of the individual spectra. If the orbital velocities of the stars are high enough, we shall observe a doubling of all the lines; in each pair the violet-shifted line is produced by star A and the red-shifted line by star B. In configuration (b) the stars are stationary relative to be observer; that is to say, the radial components of their velocities are nil. Consequently, there is no Doppler effect and no doubling of the lines. In configuration (c) we observe a Doppler effect again, but in this case the violet-shifted line in each pair is produced by B and the red-shifted line by A. In configuration (d) there is no Doppler effect and the spectrum is the same as in (b).

It is from such periodic variations in the spectrum that we can recognize a spectroscopic binary. Naturally, the actual conditions may be different. For example, if the orbital velocities of the stars

Figure 46
Illustration of the cause of
the spectral variations that
reveal the existence of a
(spectroscopic) binary system. Four typical positions
of the stars are shown; for
the intermediate positions,
(a) and (c), each star's velocity will have only a radial
component. The stars are
shown revolving about the
common center of mass, as
they actually do.

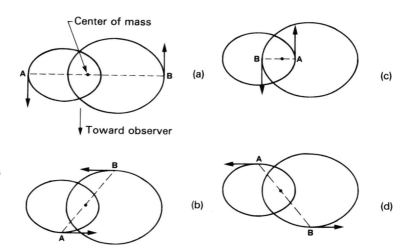

are not high enough to produce a marked Doppler effect, instead
of a periodic doubling of the lines we observe a periodic widening
of the lines. If the spectra of the binary components are not the
same, we only observe a periodic doubling or widening of the
lines common to both spectra. If one of the stars is too faint to
produce a spectrum, we observe a periodic shifting in wavelength
of the spectral lines of the brighter star. It is evident that there
can be all kinds of different combinations, but once again what
I wish to emphasize is that we can garner knowledge of the stars
by the most disparate techniques.

What can be learned from spectroscopic observations? Obviously, the velocity of one of the stars with respect to its companion—or, more precisely, the radial component of that velocity.
This time, instead of a light curve showing the brightness variations
of the binary system as a function of time, we obtain a velocity
curve. This curve must be periodic, since after a full revolution
everything starts all over. From such a curve geometrical and
temporal elements of the orbit can then be derived.

Mizar A was the first spectroscopic binary to be detected, by
E. C. Pickering in 1889. Since Mizar A was the brighter component
of a visual binary, Mizar, it followed that Mizar was a triple
system. Later on, in 1908, it was discovered that Mizar B was
also a spectroscopic binary, so that the system Mizar became
quadruple. In the same year Alcor, a small star close to Mizar,
was also found to be a spectroscopic binary, and Mizar became
a system of six stars.

Algol, too, was discovered to be a spectroscopic binary, and this finding made it possible to choose between the two hypotheses advanced a century earlier by Goodricke to explain the star's marked variations in brightness. Thus Algol is a spectroscopic binary as well as an eclipsing binary. Today we know that double stars are the norm rather than the exception; as I mentioned earlier, at least half of the stars in the sun's neighborhood are double or multiple systems.

·MORE ON DOUBLE STARS

To complete the picture, I should mention two more ways of detecting double stars. If the secondary star in a binary system is too faint to be observed, its presence can be revealed by the motion of the primary star, which instead of moving in a straight line, as a single star does, appears to pursue a wavy path. Stars of this type are called astrometric doubles because their duplicity is revealed by positional measurements typical of astrometric observations. One of the best-known stars in this class is Sirius. In 1884 Bessel described its motion as a sinuisoidal curve with a period of 50 years. Sirius's invisible companion was discovered 22 years later by A. Clark. It turned out to be a small, dimly shining, white star—the first white dwarf ever to be observed. If you recall the H-R diagram, white dwarfs lie below the main sequence to the left. Although hot (spectral classes A and F), they are very faint objects; hence they must be very small, about the size of planets. On the other hand, we know from studies of binary systems that their masses are of the order of one solar mass. The inescapable conclusion is that these stars have an exceptionally high density, on the order of 10^5 times greater than the density of water. Van Maanen's star, for example, of spectral class F8 and absolute magnitude $+14.3$, is three times more massive than the sun, but its radius is only 0.006 the sun's; its density therefore is 10^7 times the density of the sun. Since the sun's density is about the same as the density of water, this means that a cubic centimeter of the star's material would weigh 10 tons on the earth. A box of matches filled with such material would weigh about 225 tons. We have just met the first oddity of the sky. Today we know so many that white dwarfs seem almost normal.

The other way of detecting double stars that I am going to mention concerns spectroscopic binaries whose orbital planes are inclined about 90° to the line of sight. In this case the stars have no radial velocities to speak of. Although the spectrum cannot show the periodic variations due to the Doppler effect, it can still show that we are dealing with a double star. Double-star components are often stars of different spectral classes. Consequently, the spectrum we obtain, which is a mixture of the individual spectra, is rather peculiar, in that it displays lines that are characteristic, say, of a hot star as well as lines characteristic of a cool star. Since the same star cannot produce both types of lines, either the spectrum belongs to a nonexistent star, which is patently absurd (since we see it) or what we are observing are two stars so close together that they appear as one. It is a single object for the telescope, which cannot separate the images, but not for the spectrograph, which by analyzing the emitted radiation demonstrates that it is double.

Binary systems are of fundamental importance to astronomy because of the wealth of information they yield about celestial objects—and especially about stellar masses. These masses cannot always be determined directly, but accurate estimates can be made by combining the results of different types of observations. For example, in the case of eclipsing binaries that are also spectroscopic binaries, we can obtain the radii, that is, the dimensions, and the emissivities, that is, the temperatures, of the components.

· THE MASS-LUMINOSITY DIAGRAM

There are thousands of eclipsing binaries. But only a small number of them can also be observed spectroscopically, thereby enabling us to obtain the needed data. Consequently, there are relatively few double stars of which all the characteristics, masses included, are known with good accuracy. They are enough, however, for us to construct a diagram plotting their masses against their absolute magnitudes (or luminosities). Once again it turns out that most of the points do not fall at random, but on a line—or, rather, a belt; this clearly shows that there is a correlation between the two quantities. Essentially, the diagram says that the more massive the star, the higher its luminosity.

The mass-luminosity relation (see figure 47) is a consequence of the laws that govern the structure of stars. More about this later. For the time being, let us just observe that the mass-luminosity relation works well for main-sequence stars, but fails for giants and makes no sense for white dwarfs. The reason for this will be explained in part III.

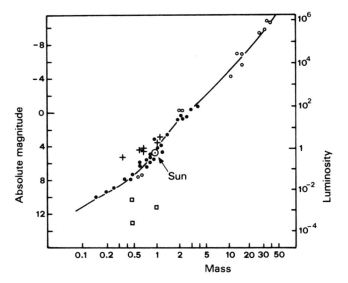

Figure 47
Mass-luminosity correlation for main-sequence stars. It shows that the two parameters are not independent; a "normal" star cannot have a small mass and high luminosity or a large mass and low luminosity. It also shows that the luminosity increases much faster than the mass. This means that the energy radiated by a massive star is more than proportional to that of a star of small mass. As a consequence, the evolution of a massive star is much more rapid than that of a star of small mass. Three white dwarfs are shown in the graph. The correlation does not work for these stars, whose "abnormality" is a consequence of the evolutionary stages in which they happen to be. Legend: •, visual binaries; ○, spectroscopic binaries; +, visual binaries of the Hyades; □, white dwarfs.

One important consequence of the mass-luminosity relation is that it can be used to determine the masses of all the single stars whose absolute magnitudes are known.

You will notice in figure 47 that the range in stellar luminosities (from a few thousandths to hundreds of thousands times the sun's luminosity) is much wider than the range in stellar masses (from a few hundredths to tens of times the solar mass). You will also notice that, roughly, the luminosity is proportional to the power 3.5 of the mass. If two stars have a mass ratio of 3, their luminosity ratio is about 45. This means that a massive star radiates much more "generously" than a light star. We shall see later on how this fact bears on the life of a star.

Stellar sizes, like luminosities, also fall within a wide range, and since the range in stellar masses is fairly limited, it follows that the range in densities must also be very wide. Main-sequence (or dwarf) stars, such as the sun, have densities ranging from one-tenth to about three times that of water. The giant stars, on the other hand, are not only much brighter, but also much larger than the dwarfs of the same spectral classes. Since their masses cannot be very different, they must have extremely low densities; for example, the average density of the supergiant Antares must be less than one-millionth that of the sun. This explains why the spectral lines of the dwarf stars are broad, while the lines of the giants are narrow and those of the supergiants are very narrow; if you recall, the higher the density of the gas, the broader the spectral lines. Thus we can now answer the question left unanswered a little while ago, namely: In which zone of the H-R diagram ought we to put a star when, knowing its spectral class, we want to determine its absolute magnitude?

·TO CATCH OUR BREATHS AND TO INTRODUCE WHAT FOLLOWS

This could be the end of the story. We have seen how astronomers have given us the measure of the sky by solving the problem of stellar distances. Today we know a lot about the specks of light that for so many centuries were a mystery to man, and along with the mystery the fear is gone because what we understand is no longer frightening. Fear has given way to a sense of wonder, the joy of understanding, and the satisfaction of having built a coherent view of the world. It may not be exact in every detail. But that does not matter. After all, we would not be ashamed if, asked to talk about a city, we could not describe in detail every street, square, house, and tree and every flower in its gardens. It is true that there is a difference between knowing a city as a point on a map and describing every part of it in every physical detail. But suppose we had such a detailed description; would we really know what the city is all about? Certainly not. We would still need to know the relations between its inhabitants, the laws that govern the city, what happens in it, its foreseeable future, its past history, its customs and curiosities. A city is a world in itself.

It is no different with the world of stars. In order to describe such a world and to make it understandable, the truly essential elements must be filtered out of the pool of knowledge built up over the years. Minute details, unless they play a key role, are best left out, particularly because they may turn out to be unrelated facts that do not add to our basic knowledge. We shall continue to explore the sky as if it were an unknown city. When necessary, we shall try to reason together. Having come this far in the book, you must have an idea of how astronomers work. In this respect, astronomy is a little different from the other sciences. We cannot cause phenomena to happen or make stars according to theories of star formation and then see whether the results agree with our assumptions; nor can we use the laboratory to test our work methods. What we do, instead, is to look at each fact, to give a reasonable explanation for it, and to fit it into an overall view of the world that is believable because it is self-consistent and constantly put to the test of experience.

The sky is full of things to see. To discuss them all, I would have to write a very large book—indeed, a whole series of large books. But we should at least take a wider look at our great city in the sky—our galaxy. Thus far we have only looked at a few things in our neighborhood, not far from home. It might be interesting to venture out into the city and learn a little (just a little) more about its inhabitants.

We have already met quite a number of stars: large and small; hot and cool; single, double, and multiple. In our census of the stellar population it is now the turn of the variable stars.

·THE VARIABLE STARS

As their name implies, these are stars whose brightness fluctuates in time. They are not to be confused with eclipsing binaries, whose variability simply depends on the occulting of one star by the other; the variability of a variable star is an intrinsic characteristic of the star. In this sense the sun is also a variable star in that its ultraviolet and x-ray emissions fluctuate with a period that is tied to the cycle of solar activity. Should the visible radiation from the sun also fluctuate (actually it does, but to a very small extent), life on earth might cease to exist. As it is, the sun's variability is harmless because, as you know, ultraviolet and x rays are absorbed by the atmosphere. By variable star we normally mean a star whose energy flux in the visible is not constant in time, that is, a star that appears more or less bright depending on the time of observation.

When first discovered, variable stars were something of a shock, at least in Europe, but in time they contributed in some measure to the development of modern astronomy. In 1572 Tycho Brahe observed a star in Cassiopeia that had never been seen before. How could something new appear in an immutable sky? The answer to this question, which could have spelled the end of many old beliefs, was to pronounce the disturbing event a miracle. Although it was a truth never to be doubted that God had created a sky that would not change for all eternity, it was also true that God in His omnipotence could change His mind at any time. Earlier on, in 1504, the Chinese and Japanese had observed a new star in the constellation Taurus that was visible in broad daylight. In the Mediterranean basin the star was not noticed, but it was seen by Navaho Indians in North America, as shown by two rock carvings discovered in northern Arizona. Another new star was observed in 1604 by Kepler and others in the constellation Ophiuchus. All these stars, and others of the same type, were called *novae*. More about them later on.

People get used to everything. And when miracles happen a bit too frequently they are demoted to the rank of phenomena. As such, they are no longer the concern of priests and become the exclusive property of the scientists. This is what happened with novae. In the past, however, novae and variable stars in general were not investigated extensively. The hunt for variables is recent history, mainly because photography is a recent development. It is practically impossible to look for variables by eye, even if armed with binoculars or a telescope. Finding one is a rare chance, particularly because the only variables that can be discovered by eye are those that exhibit very conspicuous and very rapid variations in brightness. This is the reason why only ten or so variable stars were known by the beginning of the nineteenth century. Photography has made all the difference. The way to find variables is to photograph a star field many nights in a row and to compare the plates. There are special instruments that enable us to observe pairs of plates in rapid succession; since images persist in the eye, stars that have not fluctuated in brightness appear the same throughout, while the variables "scintillate" and therefore can be easily identified. Once this operation is completed, additional research into these stars can be carried out with photoelectric and spectroscopic techniques.

There are different kinds of variable stars, but to begin with, a distinction can be made, based simply on general descriptive characteristics, between two types of variables whose behavior is radically different—regular or periodic variables and irregular variables. The former exhibit regular oscillations in brightness about an average value, while the latter fluctuate in a more or less unpredictable fashion.

·REGULAR VARIABLES

Let us consider the distinguishing characteristics of the 10,000 or so regular variables currently known. Several classes have been established based on the shape of their light curves (which show light variations as a function of time) and their periods of oscillation (from a few minutes to 100 days). The best-known group is that of the Cepheids, which are very regular variables with periods ranging from 1.5 to 50 days (see figure 48 for a typical light curve). About a thousand are known. Many of them were discovered in the Milky Way, and it is now well established that these stars are located on or near the galactic plane. I have never mentioned this plane, and I would rather not talk about it right now, but I dislike making meaningless statements. I shall just say that all the stars we see in the sky belong to our great island in space—our galaxy—which, seen from the side, is shaped like a disk with a bulge in the middle, while seen from above, it looks like a spiral with a central nucleus and spiral arms wrapped around it. All around our galaxy, almost like satellites, there are about a hundred minor stellar systems called globular clusters. The plane that cuts our galaxy in two edgewise, the way we would cut an English muffin, is the galactic equatorial plane.

As I was saying, the Cepheids lie on or close to this plane and are situated at the periphery of the disk, in the spiral arms. They are stars of low velocity. During each light cycle a Cepheid exhibits conspicuous changes in color, that is, spectral class. Spectroscopic observations show that the spectral lines undergo displacements that are in perfect agreement with the Doppler-effect relation. Such displacements occur periodically and with the same period as the light variation. The light curve, showing variations in magnitude—that is, luminosity, since the star cannot possibly be alternatively approaching and receding from us in a manner affecting its magnitude—and the radial-velocity curve obtained from the spectra are in phase, with maximum light occurring at the same time as the maximum velocity of approach (see figure 49).

The accepted interpretation of this phenomenon is the following. A Cepheid is a pulsating star, that is, a star that at regular intervals expands gradually to a maximum size, then contracts to a minimum size, then expands again, then contracts, and so on over and over again. Quite a surprise. But the pulsation theory fits the facts, and it is very hard to find a more satisfactory explanation. The greatest expansion velocity corresponds to the greatest value of the red shift and to the passage (in the expansion phase) through the mean value of the diameter. At this point the temperature and luminosity are at a maximum. When the star reaches its maximum size, the expansion velocity is nil and the temperature and luminosity have mean values. Then the star begins to contract. When it reaches its average size, the contraction velocity is at a

Figure 48
Schematic light curve of a
Cepheid variable. The light
variations are very regular in
both amplitude and period.

Figure 49
Part (a), the radial-velocity
curve, shows the variations
in the radial velocity of a
Cepheid in the course of a
pulsation. The broken line
indicates the star's own radial velocity (excluding the
pulsation). Part (b), the light
curve, shows the same
Cepheid's light variation in
the same period.

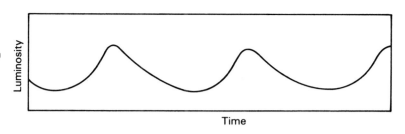

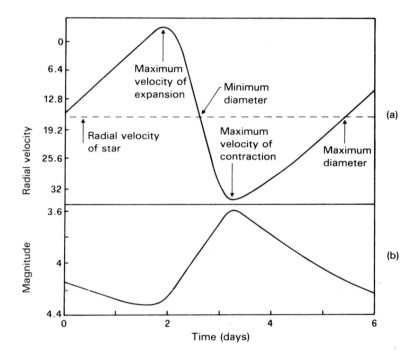

maximum and the temperature and luminosity at a minimum. Finally, the star contracts to its minimum size, the contraction velocity is nil, and the temperature and luminosity have mean values. And then the cycle starts all over again.

A simple calculation will give us an idea of the amplitude of the pulsation in kilometers. All we have to do is to multiply the average expansion velocity by the time the expansion lasts. In actuality the calculation is a little more complicated, but it does not matter because the orders of magnitude are good enough for our purpose, which is to understand how we can find out things about objects that are very remote and appear as no more than tiny specks of light. In the case of δ Cephei, for example, we find that the distance covered by the star's photosphere between the phase of maximum contraction and that of maximum expansion is approximately 2.5×10^6 km. From the star's spectral class (that is, temperature) and absolute magnitude, we can derive its radius, which is about 40×10^6 km. This means that a pulsation corresponds to about a 6% variation in the size of this star.

The most important property of the Cepheid variables is the correlation between their periods and intrinsic luminosities: the longer the period, the higher the average luminosity. For example, of two stars whose periods are 30 days and 3 days, respectively, the former is about six times brighter (a difference of two magnitudes). The correlation was first noted by H. Leavitt from her studies of Cepheid variables in the Small Magellanic Cloud. (The Magellanic Clouds are two dwarf galaxies—10 billion stars each—that are satellites of our own.) Since in 1912 the distance of those Cepheids was not known, it follows that the correlation was based on the apparent magnitudes rather than the absolute magnitudes. However, although close by as galaxies go, the Small Magellanic Cloud is far enough away that all stars in it can be regarded as being at the same distance from us. There is nothing wrong with this line of reasoning. You too would say that all the passengers on a plane in flight, or a ship at sea, are at the same distance from a ground observer. But if the Cepheids in the Small Magellanic Cloud are all at the same distance from us, apparent magnitudes will give the same information as the absolute magnitudes; that is to say, a star that looks three times brighter is intrinsically three times brighter. Since in 1912 the distance of the Magellanic Clouds was not known, the absolute magnitudes of the Cepheids could not be determined. (This had to wait for the determination of a certain constant.) What they knew, therefore, was the *difference* in magnitude between Cepheids of different periods.

Harlow Shapley understood that the Cepheids could be of fundamental importance as distance indicators. The reasoning goes like this. Let us assume that the Cepheids are all alike throughout the universe. If we can calculate the distance of one of them (or better, a few of them), we can find the relation between its period of light pulsation and its absolute magnitude. From this relation and the assumption, we can estimate the absolute magnitude of any Cepheid of known period, no matter how far away. From the absolute and apparent magnitudes we can then derive the parallax, that is, the distance of the Cepheid, as well as the distance of the stellar system to which it belongs. This is another way of determining celestial distances, incidentally, and it was research along these lines that revealed the universe to be even larger than had been thought.

This explains the importance of the period-luminosity relation. But in order to use it we must know the intrinsic luminosity of at least one Cepheid, or, as we say in technical jargon, we must establish the zero point of the period-luminosity law. Unfortunately, there is no Cepheid close enough for us to measure its distance by the trigonometric method. But no astronomer can have a good day's sleep (at night he generally works) until a way is found of determining the distance of all celestial objects, which is an essential piece of information to obtaining the intrinsic luminosity. In the case of the Cepheids reliance had to placed on statistical methods based on the proper motions and radial velocities of the stars. This forces upon me another brief digression.

•PROPER MOTIONS AND STATISTICAL PARALLAXES

At this point I do not think anybody still believes that the stars are "fixed," as the ancients thought. Indeed they move, and even though each star goes its own way, they all share in a general motion around the center of mass of our galaxy.

The true motion of a star is called its proper motion. This is not to be confused with its apparent motion, which is an optical effect due to the earth's motion around the sun; it is a star's apparent motion that produces the displacement used to measure its parallax. Because of its proper motion, each star's position changes with the passage of time; the result is that in time the disposition of the stars on the celestial vault becomes different from what it used to be. Figure 50 illustrates the change in time of the relative positions of the seven main stars in Ursa Major. It takes a very long time to observe changes in the sky. This is because although the stars travel at fairly high velocities, they are exceedingly far away, and therefore our lines of sight to them move very slowly. Suppose you are on the moon and move 400 km in one direction. For an observer on the earth, the angle between the line of sight to where you were and the line of sight to where you have moved is only about 3' (3/60 degree).[46] And the moon is only 400,000 km away, or a little more than a light-second (the distance that light travels in a second of time). Suppose now you are on a star 50 light-years from earth and moving with a transverse velocity (that is, at right angles to he line of sight) of 100 km/sec. In a year the star travels about 3×10^9 km. A simple calculation shows that 50 light-years (about 15 parsecs) are about 5×10^{14} km. Thus the angle formed by two lines of sight from the earth to the star, of which one is taken at the beginning of the year and the other at its end (thereby eliminating apparent motion, since a year's time restores the earth to its original position with respect to the sun), that is, the star's proper motion, is about 1" (1/3,600 degree). Try it if you do not believe me. Such small displacements cannot possibly be detected with the naked eye, and it is quite understandable that the ancients should have believed that the stars stood still.

The proper motion of a star is thus the annual displacement, expressed in seconds of arc, of our line of sight to it. The displacement is generally smaller than the one I have given as an example,[47] and therefore it takes many years to measure proper motions with good accuracy. The current method consists in com-

Figure 50
The seven main stars in Ursa Major at different times. The change in the relative positions of the stars is due to their proper motions.

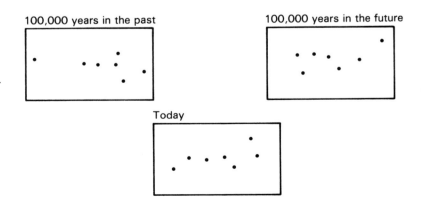

100,000 years in the past

100,000 years in the future

Today

paring photographs of the same star fields taken over an interval of several years. Most of the stars in these photographs do not change positions because they are so far away that it takes a long time for any movement to be appreciable. But some stars do move in relation to the other stars. These are the stars closest to us, of course, or the very few stars that are extremely fast moving. The greatest proper motion (10.3″/year) belongs to Barnard's star (figure 51), but it is only partially due to high velocity; at a distance of 1.8 parsecs, Barnard's star is the star second nearest to us after α Centauri.

Let us turn now to a star's space velocity, which is the velocity of the star (in km/sec) in relation to the sun. It is convenient to divide it into two mutually perpendicular velocities. In figure 52,[48] v represents the annual motion of star S, v_r is the component of its space velocity in the line of sight (radial velocity), and v_t is the component of its space velocity at right angles to the line of sight (tangential velocity). If SS′ is the annual displacement of the star on the sky, then μ is the star's proper motion. The radial velocity is independent of the distance and can be obtained from measurements of the Doppler shift. The tangential velocity is given (in km/sec) by

$$v_t = 4.74(\mu/p),$$

where p is the parallax and μ the proper motion.[49]

Figure 52 actually refers to an observer on the earth, but it is not difficult, at least in principle, to refer a star's motion to an observer on the sun, that is, to give the proper motion and the components of the space velocity in relation to the sun. To do this, one must take into account the effects of the earth's rotation and orbital motion on the quantities measured.

As far as radial velocity is concerned, there is not much to add to what we have already learned. Given a spectrum of sufficient dispersion, it can be measured from the shift of the spectral lines due to the Doppler effect. The radial velocities of 25,000 stars have been measured to date—not very many, but you must remember that it is hard to obtain suitable spectra.

Things are more complicated in the case of tangential velocities. What we measure—the proper motion—is an angle that represents the displacement of a star on the celestial vault. It is obvious that for a given value of v this displacement is the smaller the farther the star; hence the proper motions of very distant stars can only be measured after long intervals of time. To date, proper motions

Figure 51
Barnard's star (indicated by
arrow). Note that while the
relative distances of the
field stars have remained
the same in the two photographs, the position of Barnard's star has changed
markedly in only 22 years
as a result of its large
proper motion.

Figure 52
The observer is at O, the
star S is at distance d, and
μ is the star's proper motion; v is the space velocity
of the star; v_t and v_r are, respectively, the tangential
and radial components of v.
The proper motion depends
on the tangential velocity; if
v_t is nil, the line of sight OS
does not change over time
and the position of the star
on the celestial vault remains the same. (Its motion
along the line of sight could
be deduced from the shift
of its spectral lines due to
the Doppler effect.)

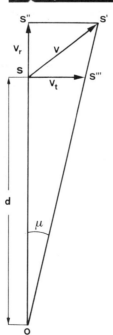

have been measured for about 300,000 stars that are fairly close, within 200 parsecs of the sun. Since the tangential component of a star's space velocity depends on the star's proper motion and distance, and since the possibility of measuring the proper motion also depends on the distance, that is, the parallax, proper motions can be used to estimate distances. Let us see how it can be done, at least in general terms.

It is reasonable to assume that the stars do not have the same velocity. But, proceeding statistically, we can always select a group at random and calculate its average space velocity. If we do the same with different groups, of 50, 100, or 150 stars, it is also reasonable to expect that we shall find approximately the same value for each group. While this is true for the average space velocity, it is not true for the average proper motion, which depends also on distance. (It is inversely proportional to the distance.) Thus we can assume that the size of the average proper motion of a random group of stars is a good indication of that group's average distance. From a group's average proper motion, that is, average distance, and from its average apparent magnitude, we can determine its average absolute magnitude, which is one of the fundamental quantities we need to know. To give an example, we can take a particular group of stars chosen on the basis of some characteristic, such as magnitude or spectral type (but not proper motion, either directly or indirectly, or we would affect the average value we are seeking), and determine the average distance of the group, that is, the average parallax. Parallaxes obtained by statistical methods are called statistical parallaxes. Since proper motions produce displacements that increase with time, this method can be used to estimate the parallaxes of stars that are too far for us to use the trigonometric method. Such is the case with the Cepheids and RR Lirae stars, none of which are close enough for direct measurements.

·BACK TO REGULAR VARIABLES

Early estimates of the absolute magnitudes of the Cepheids were not very accurate because in the spiral arms, where these stars are located, there are dust clouds that dim the light to an extent that is hard to evaluate; this makes the determination of the apparent magnitudes very difficult. Although current measurements are still not as accurate as one might wish, the results we have today are pratically certain. Galactic, or Population I[50], Cepheids with periods slightly longer than 1 day have an absolute magnitude of about −1.5, while Cepheids with a period of 100 days have an absolute magnitude of about −6. The others fall between these two outside values more or less on a straight line (see figure 53).

The search for Cepheids, or pulsating stars, produced two other main groups: the W Virginis stars and the RR Lyrae stars. (Variable stars are named after the prototype of the class; classical Cepheids are also known as δ Cephei stars).

The W Virginis stars (or Population II Cepheids) are found in globular clusters and in the galactic corona. Their light curves are similar to those of the classical Cepheids, but the periods are long (10 to 30 days), and the decline from maximum light is slower. Also, these stars are bluer in color. A period-luminosity law seems to apply to these variables as well, in the sense that a longer period corresponds to a higher luminosity.

The RR Lyrae stars are the most common variables in globular clusters. They are also found in the nucleus of our galaxy and in the galactic corona. Their periods are very short (less than a day), and their light fluctuations never exceed two magnitudes (figure 54). RR Lyrae stars fall within a small range of spectral classes, from A to F.

Figure 53
Period-luminosity relation for Cepheids and RR Lyrae stars.

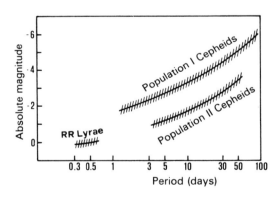

Figure 54
Light curve of an RR Lyrae variable showing the variations in magnitude. The oscillations are regular and of short period.

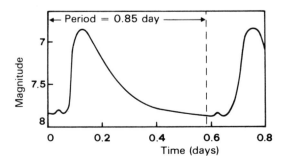

If we estimate the average apparent magnitudes of the RR Lyrae stars in any globular cluster, we find that they are approximately the same. Since the stars in a cluster can be regarded as being at the same distance from us, it follows that they also have the same absolute magnitude. And since the RR Lyrae stars in other clusters have the same characteristics, it is reasonable to conclude that all RR Lyrae variables have the same absolute magnitude. In other words, we can assume that a star that happens to be an RR Lyrae has acquired a set of characteristics that results in a specific luminosity. Consequently, the RR Lyrae stars are as important as the classical Cepheids as distance indicators; actually they are even better, in that the absolute magnitude of any RR Lyrae star is known a priori (it is the same for all of them). Thus if we observe one of these stars and determine its apparent magnitude, we can find its distance. The problem is that we have to measure the distance of at least one of them to obtain the absolute magnitude once and for all. Once we have determined the absolute magnitude, proceeding backward we can find the distance of any stellar system that contains RR Lyrae stars. As with the Cepheids, however, no RR Lyrae stars are close enough for us to measure trigonometric parallaxes, and once again reliance has to be placed on less direct statistical methods. The absolute magnitude of these stars turns out to be 0—or, more precisely, between 0 and +1; that is, it can vary slightly from one star to the other.

Other variables worth mentioning are the Mira Ceti stars, cool red supergiants with periods up to 700 days and light variations in excess of 2.5 magnitudes; the δ Scuti stars, whose periods are very short (a few hours) and whose light variations range from 5/100 to 5/10 of a magnitude; and the dwarf Cepheids, whose periods are also quite short. There are others, but I do not see the point of running on with a systematic classification when my purpose is to give a general view of our galaxy rather than a plethora of information that in the long run would only be a distracting element.

· IRREGULAR VARIABLES

Let us move on to the irregular variables, some of which are important in the exploration of our galaxy. In general, an object is described as "irregular," which usually stands for "likely to behave in a strange, unpredictable way," because it is not yet fully understood. In some cases the causes of the irregular behavior are understood and the object is less mysterious, but it is still very hard to grasp the dynamics of the phenomena that produce the irregular behavior. The first step in seeking to understand irregular phenomena is to group them on the basis of similarity; then one joins theory and observation in order to explain them. Sometimes it becomes evident that they are the outcome of a whole series of events and it is even possible to trace the evolutionary history of which they are one episode. In some cases, however, the rarity of the phenomenon and the almost total absence of any ties with other known phenomena make understanding all but impossible. The psychological difficulty of living in uncertainty, combined with mental laziness and prejudices, often leads people to give up the search for a rational explanation and to turn, instead, to talk of miracles. It does not happen so much today, but it was not too long ago that miracles were invoked to explain events like eclipses and comets. Fortunately, this habit has fallen into disuse, at least with most of us.

Irregular variables are stars of strange behavior—either unpredictable or predictable but irregular—which will not be fully understood until much more is learned about them. I could make an analogy with mental patients. Once they were thought to be possessed by the devil; then they were thought to be "crazy"; and now they are simply "sick." And the more they are regarded and treated as sick people, the more can be learned about the causes of their abnormal behavior. Of course, I do not mean to imply that a variable star is a sick star. Analogies should not be carried too far, and I wish to stress this point because all too often people fall in love with their own images and embroider on them until the original concept is lost. No matter how apt it is, an analogy is not the real thing and should be taken with a grain of salt.

These stars may be grouped into three main classes: variables of the R Coronae Borealis type and flare stars; novae and supernovae; and the intermediate group of U Geminorum and Z Camelopardalis stars.

Briefly, an R Coronae Borealis variable is a star whose luminosity remains fairly constant until suddenly the star dims by 4 or 5 magnitudes (see figure 55). It takes hundreds of days for the star to recover its original brightness. So far, no regularities have been found in either the amplitude of the light variation or the time interval between one dimming and the next. In this sense such a star is unpredictable. But we know that a star that has gone

Figure 55
Light curve of the variable
star R Coronae Borealis
showing the variations in
magnitude. As you can see,
it shows no apparent
regularity.

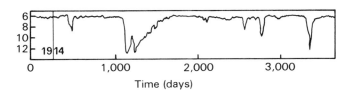

through one such phase (that is, a variable of the R Coronae
Borealis type) will go through other similar phases; and in this
sense such a star *is* predictable and hence classifiable. At maximum
light stars of this type resemble supergiants of spectral class F.

Flare stars are dwarf stars of spectral classes F5 through G5 (T
Tauri stars) and cooler (UV Ceti type) that are often associated
with nebulosities. In the H-R diagram they lie just above the main
sequence. Their behavior is wholly erratic. At times they brighten,
and during these phases they display irregular light fluctuations
that, though on a very different scale, call to mind the phenomena
of solar activity. At such times, if the interpretation is correct,
these stars must emit a considerable flux of particles, just as the
sun does during a solar flare. At maximum light they show emission lines and radiate prevalently in the ultraviolet range of the
spectrum.

Novae and supernovae, needless to say, are not brand-new
stars (*nova* is Latin for new) come to life as if by magic, but stars
already in existence that pass through a catastrophic phase. A
nova is a normal star that has exploded and becomes very bright.
Before the explosion a nova is too faint to be noticed, so that for
us it is as if a new star had appeared in the sky; after the explosion,
however, it can be traced back to the original object in old plates
of stellar fields. Suppose it was a star of magnitude 10, which
certainly cannot be seen with the naked eye. Suddenly, where
there seemed to be nothing, a star appears (say, a star of magnitude
6) that in the space of a few days becomes brighter and brighter
until it rivals the brightest stars in the sky. The initial stages of
the event are so rapid that they may go unobserved. All it takes
is a week of bad weather. Or a failure to take notice. As to the
last point, you must know that the observational programs of
astronomical observatories are planned and approved months in
advance—the better equipped the observatory, the farther ahead
the plans—and that researchers take turns using the observatories'
large instruments according to precise schedules. No astronomer
can simply walk in and take a look at the sky whenever the spirit

moves him. A large telescope costs millions of dollars; obviously, it is not a toy. Contrary to popular belief, therefore, astronomers *never* spend the night at the telescope just to enjoy the beauty of the starry sky and to share their emotions with like-minded colleagues. Like bus drivers and telephone operators, if it is not their turn and if they do not have a specific job to do, they do not hang around.

Under these conditions, it is very unlikely that an observer will chance upon a nova explosion. As in the case of comets, novae are often discovered by amateur astronomers, whose love for the sky is certainly greater than that of the professionals, although perhaps more irrational (I did not say unreasonable), in that it is motivated by esthetic rather than scientific considerations. Unfortunately, with the limited equipment at his disposal, the amateur cannot do much more than alert the astronomical community to the appearance of a nova. In most observatories the news causes something of a commotion because a nova explosion is such a rare event that no one can afford to miss it. Programs have to be scrambled or cancelled to the dismay of the interested parties. Although the people who are bumped off the telescope cannot help wishing that the darn thing had picked another time to go off, everybody agrees that it is a wonderful opportunity for science. The thought that the new data might benefit a colleague rather than oneself never enters one's mind. Clearly, science comes first.

If it is hard to observe the course of the explosion, it is even harder to collect much information concerning the prenova stage. It only happens by chance, while observing other objects, because at present we have no way of telling which stars are going to explode—we have only our suspicions. In general, prenovae are faint, fairly hot stars of spectral class A, but there are lots of stars in the sky that fit this description. It would be like searching for a Scandinavian writer about whom we know only hair color (blond) and color of eyes (blue); obviously, a better description is needed. It is the same with prenovae. Perhaps when more data are available, a number of common characteristics will emerge and many questions will be answered.

A few hundred novae are known well enough that the phenomenon can be sketched in general lines, and we shall see later on that a satisfactory explanation has, indeed, been found for it. We owe much of our knowledge of it to spectroscopic observations, and this demonstrates once again the fundamental role this instrument plays in the exploration of the universe.

Let us summarize the essential features of this phenomenon. In the course of the outburst the spectrum of a nova resembles that of a supergiant of spectral type A. The spectral lines are shifted to the violet, however, which indicates that the gas is approaching the observer; that is, the star is expanding. From the magnitude of the Doppler shift, we can deduce the radial velocity of the source (the expanding gas in this case), which turns out to be on the order of 1,000 km/sec. More than an expansion it is an explosion, and quite an explosion at that. Furthermore, the bright emission lines appear that are characteristic of nebulosities. The inescapable conclusion is that we are observing a cloud of gas that is rapidly expanding and leaving the star. It is a cloud of considerable proportions. Although our estimates are not too accurate, it appears that in the course of the explosion the star loses a mass equivalent to many times the earth's mass. The density of the particles in the ejected shell can be inferred by the intensity of the lines it produces. (Clearly, there must be a relation between the number of atoms that produce spectral lines and the intensity of the lines.) Knowing the density, even roughly, and the dimensions of the shell (which sometimes becomes visible on photographic plates), we can estimate the total number of atoms in the shell and hence its mass.

The explosion, as can well be imagined, entails a considerable expenditure of energy. And more energy is expended during the months in which the star, though becoming increasingly fainter and eventually subsiding to the original level of luminosity, remains brighter than normal (figure 56). The total energy that a nova expends during an outburst has been estimated at about 10^{45} ergs. At its current rate of emission it takes the sun 10,000 years to radiate the same amount of energy that a nova dissipates in a few months. It is clearly a stupendous event. If something of the kind happened to the sun, as far as we are concerned it would be the end of the world. Imagine the sun swelling like a balloon and spreading over more and more of the sky, which would become as bright as or brighter than the sun itself. In the end there would be no more sky because we would be inside the expanding cloud. We could not see any of this, of course, since every living thing on earth or underwater would be dead.

Nevertheless, for a star to become a nova is not such a catastrophe. For the sun a mass loss of the order of the earth's mass is totally insignificant, considering that the earth's mass is about three-millionths that of the sun. It is as if a 70-kg man lost 0.2g.

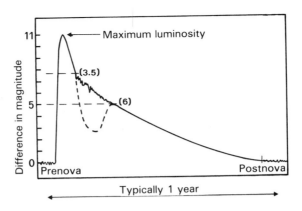

Figure 56
Schematic light curve of a nova. The rise to maximum luminosity is very rapid, and the star may brighten by as much as 11 magnitudes. Following maximum luminosity there is fairly rapid decline in brightness. After the star has faded by 3.5 magnitudes, there is a period characterized by either a series of pulsations or another steep decline in brightness followed by a rise. When the brightness has fallen by 6 magnitudes below maximum, the nova enters a fairly placid period and slowly fades away to its original brightness.

You lose more weight by trimming your nails, and 50 times more weight when you give a blood sample for medical tests. And, in fact, when the explosion is over a nova slowly returns to its original condition as if nothing had happened. What has happened, actually, is that the star has ejected its outer layers. We shall see later how this phenomenon is explained. But in any case there is nothing for us to worry about. If our theories are correct, and there seems to be no reason to doubt them, the sun is in no danger of becoming a nova.

Current research has added new information. Infrared observations suggest that there may be dust in the gas shell expelled by the nova. Since the dust acts as an absorbing screen, the nova may be even brighter than had been thought. Such details may affect our estimates of the energies involved, but almost certainly they will not change the general picture.

The photographs in figure 57 show Nova Herculis 1934 during and after its outburst. I should point out that the white disk on the left-hand side is due to the increase in luminosity that "burned" the plate. At that time the star was still a point source, though very bright. If the exposure time had been correct for the nova, it would have been too short for the other stars to show up on the plate.

Some novae are recurrent, but the time intervals between outbursts are irregular and very long (tens of years). Finally, most novae are members of close binary systems, and this seems to be true for all novae, even those whose duplicity has not been ascertained because of the difficulties intrinsic to observation.

And now, the supernovae. (I feel like a circus master introducing the biggest attraction in the show!) Supernovae are a hundred thousand, a million, a billion times brighter than novae. A nova

Figure 57
Two photographs of Nova Herculis 1934 showing the remarkable decline in brightness that took place between March 10, 1935 (top), when the nova was at the peak of its splendor, and May 6, 1935 (bottom), when the star had faded to almost its normal luminosity.

becomes thousands of times brighter than it was originally. A supernova explosion is a cataclysm without compare; at maximum light it is millions, billions of times brighter than before. The energy emitted is on the order of 10^{50} ergs, which is equivalent to the energy the sun would emit in a billion years. The absolute magnitude of a nova at maximum ranges from -6 to -9, that of a supernova from -14 to -20. To put it another way, supernovae attain luminosities comparable with that of the galaxies in which they appear.

Three supernovae have been observed in our galaxy in the last thousand years—in 1054, 1572, and 1604. Obviously, they were only observed with the naked eye. They are rare events, but not that infrequent, relatively speaking, because with modern instruments many of these objects have been observed in other galaxies. In a normal galaxy—a galaxy like ours, so that we understand each other—a supernova explodes, on the average, every two or three centuries. Because of their rarity we do not know very much about them or fully understand what makes a star become a supernova. The current interpretation will be discussed in part III, on stellar evolution, but for the moment let us see what little we do know.

Apart from the much greater luminosity, the light curve of a supernova resembles that of a nova. There seem to be different kinds of supernovae, but so far a twofold distinction has been made. The brightest supernovae are classified as type I; type II supernovae do not attain the same luminosity as type I, and their light curves show a more irregular decline in brightness following maximum light.[51] Each type reaches maximum in a matter of days. Spectroscopic observations show that supernovae, like novae, eject material. But while a nova explosion involves only the outer layers of the star, a supernova explosion involves a good portion of the stellar body. The velocity of ejection is on the order of 10,000 km/sec.

The supernova of 1054—or, more precisely, what is left of it—has been identified with the Crab nebula in the constellation Taurus. This much photographed nebula, which is invisible to the naked eye, consists of an amorphous gaseous mass overlaid with a network of intricate filaments (see figure 58). After the sun it is probably the best-known and most studied object in the sky, and as we shall see there are very good reasons for this. Photographs taken at yearly intervals show that the outer layers of the nebula are moving away from the center, with an angular

Figure 58
The Crab nebula, a gaseous
remnant of the supernova
of 1054.

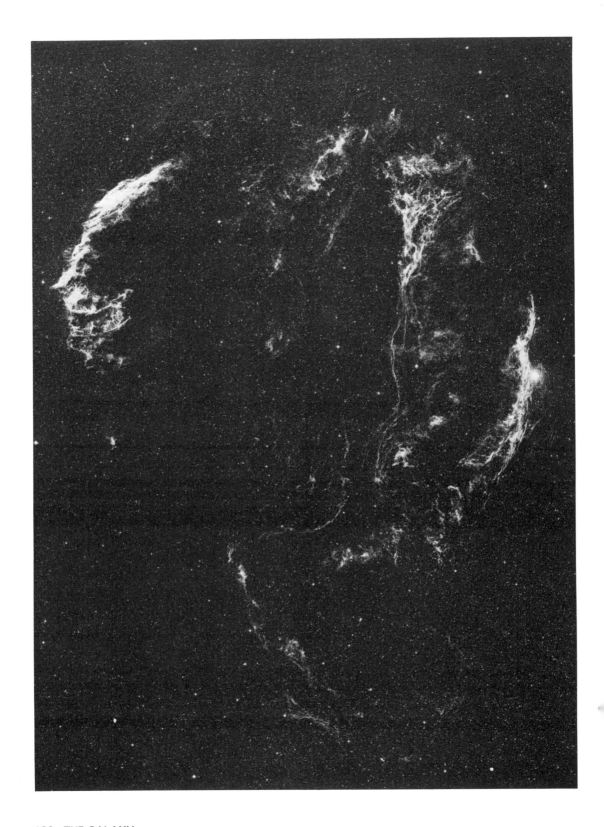

Figure 59
The Veil nebula in Cygnus,
relic of a supernova that
exploded tens of thousands
of years ago.

displacement of 0.21"/year; spectroscopic observations have yielded radial velocities on the order of 2,000 km/sec. These data clearly indicate that the nebula is expanding and enable us to estimate its distance at 6,500 light-years. If we assume that the rate of expansion has always been the same and if we work our way backward from the present radius of the nebula, which is roughly 180" (that is, if we divide 180" by 0.21"), we find that the nebula shrinks back to a point approximately 860 years ago, which brings us very close to the time the 1054 supernova was observed. The error is due to the fact that ours is but a rough calculation; in reality, the radial velocity is not the same for all points of nebula, and the nebula is not spherical but elliptical in shape. Time reckoning and the records left by the Chinese and the Navaho Indians all point to the same object. Thus there is no doubt that the Crab nebula is the relic of the 1054 supernova.

Various other nebulosities are believed to be the remnants of supernovae that exploded in a distant past; the Veil nebula in Cygnus, for example, might be all that is left of a star that blew up 8,000 years ago (figures 59–62). But the Crab nebula is still the most interesting as well as the most spectacular of all supernova remnants.

In 1948 it was discovered that the nebula is a strong source of radio emissions. At this point in the book you are not likely to believe that radio signals from the sky necessarily mean that intelligent beings out there are broadcasting frantic calls for help. Let us leave these fantasies to science-fiction writers and look at the facts. Radio emissions, like light or x-ray emissions, are electromagnetic waves and nothing more. As we learned earlier, a heated object emits electromagnetic radiation at all wavelengths. If the object is a blackbody, the emission at radio wavelengths is in a certain ratio to the emission at, say, visible wavelengths, and this ratio is provided by Planck's law (which gives the spectral distribution of the energy emitted by the blackbody). If the object is not a blackbody, the emission ratio will be different, but this does not mean that emission will not be there. Thus it is not surprising to find that an object emits radio waves. On the contrary. At low temperatures there is more emission at the longer wavelengths, and therefore it is *more natural* that at low temperatures an object should emit radio rather than light waves. As the temperature rises, maximum emission shifts to shorter wavelengths, and eventually the emission at short wavelengths becomes dom-

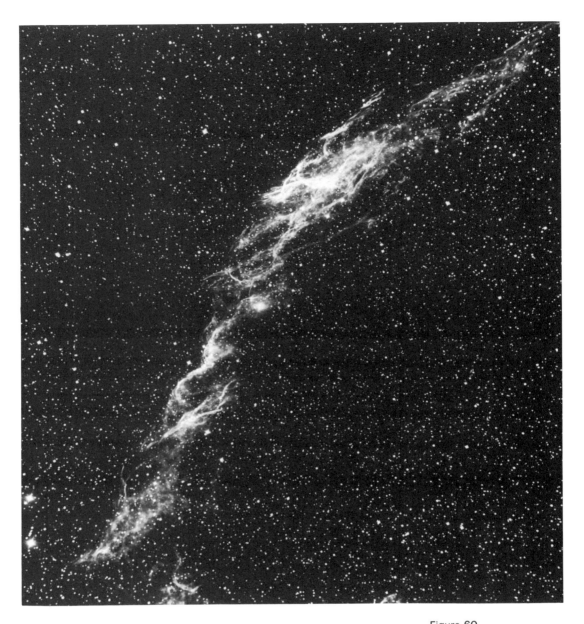

Figure 60
Detail of the Veil nebula in
Cygnus.

Figure 61
Detail of the Veil nebula in
Cygnus.

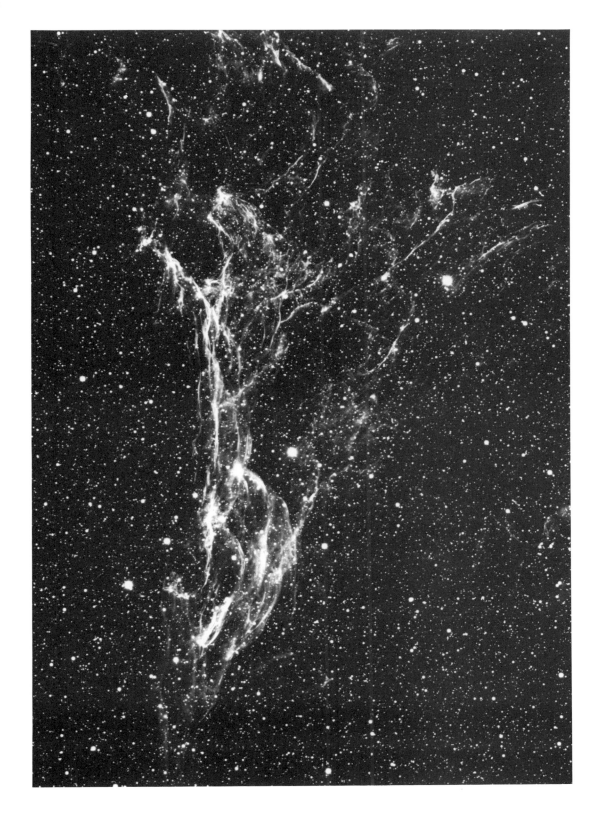

Figure 62
Detail of the Veil nebula in
Cygnus.

inant; nevertheless, the emission at long wavelengths is still there and can even be more intense than that of cooler sources. Figure 63 is a photograph of a radiotelescope, a device for detecting emissions at radio wavelengths.

I hope I have made my point. There is nothing strange about radio emission and certainly no need to fabricate fantastic explanations. No need at all, now or at any time, as a matter of fact; not even when confronted with the strangest, most extraordinary phenomenon. If we want to remain within the confines of what is called the scientific method, we must use our knowledge to find an explanation that fits the facts. Sometimes it may be necessary to modify the accepted schemes and even introduce new concepts or new laws. But then, you will ask, what is the difference between a so-called scientific fact, which is only an interpretation, after all, and always susceptible to change, and a gratuitous, unscientific statement? Basically, the difference is that a scientific prediction is substantiated by observation. The scientific method is such because it is sustained by thought processes that are foreign to pure, simple, unbridled imagination (although I am the first to say that imagination is also needed to do science). As I said before, science is concerned with the interactions between phenomena. There is no such thing as an explanation of a single unrelated fact. Thus any model we make to explain one thing has certain consequences for other phenomena. For example, if I believe that a certain phenomenon is due to magnetic fields, I must figure out what type of magnetic field will explain it. No matter how I define it, however, practically no magnetic field has only one consequence, namely, the phenomenon in question. Hence I must predict and verify all other possible consequences. A lot of facts must be taken into account, and if in the end the facts do not square, my model is only good for the wastepaper basket.

Coming back to the Crab nebula, the problem is not that there should be radio emission from it, but that the emission should be much more intense than expected. A related question, incidentally, is whether other strong radio sources could also be supernova remnants. In any event, an explanation must be found for the unexpected intensity of the radio emission. It turns out that the nebula's radio spectrum (the intensity distribution of radio waves as a function of wavelength) is similar to the spectrum that can be expected from clouds of electrons moving at nearly the speed of light in helicoidal (spiral-shaped) orbits about the lines

Figure 63
Observations at radio wavelengths are performed with
radiotelescopes, which
come in a variety of sizes
and shapes. Shown here is
the east-west arm of the
radiotelescope at Medicina
(Bologna), known as *Croce
del Nord*. It is the largest
radiotelescope in Italy.

of force of a magnetic field. Under these conditions, physics tells us, the electrons emit radiation at the expense of their kinetic energy. It is known a synchrotron radiation because it is artificially produced in synchrotrons, which, as you may know, are devices for producing high-speed electrons for research in nuclear physics. Fine. Now we make some calculations. We need a magnetic field made in such and such a way, of such and such intensity, and we need so many electrons. And there is our explanation. If it is correct, however, we should find that part of the optical emission is also due to synchrotron radiation; furthermore, it must be polarized. Observations carried out at Mount Palomar in 1956 showed that the light from the Crab nebula was indeed polarized. Our hypothesis is proved correct. Today synchrotron radiation is the accepted interpretation for both radio and optical emissions from the Crab nebula. Of course, should new evidence come to light that invalidates the model, we would have to look for another explanation.

I realize that I have worked myself into a corner. To explain one thing I have had to introduce two things with which you may not be familiar—the lines of force of a magnetic field and polarized light. The first I leave up to you to investigate; there are plenty of physics books that can explain it quite clearly. The second, instead, I will do my best to explain as simply as I can.

•POLARIZED LIGHT

God only knows what radiation is, light in particular, but we often find it useful to describe it as a wave phenomenon. This is why we all speak of light waves and radio waves, and why radio announcers tell you to turn to a particular wavelength if you wish to hear your favorite program.

The light waves, and in general all the electromagnetic waves, emitted by a natural source do not lie in any particular plane, but in all of the infinite number of planes that contain the source (think of it as a point). To understand what I mean, hold a string at one end, attach it to something at the other end so it will not move, and shake your hand. A wave propagates along the string. Consider two cases. First, you shake your hand strictly up and down; then the wave will lie in only one (vertical) plane. Second, you shake your hand at random, instead; then the vibration coming from your end of the string will lie, at each successive instant, in a different plane. In the first case the wave is said to be polarized; in the second case, unpolarized.

Suppose now you place a screen with a vertical slit between the two ends of the string and shake your hand at random (see figure 64). The slit will only let through that portion of the vibration that takes place along its length. Before the slit the wave is unpolarized; past the slit it is polarized. The screen in this case acts as a polarizer. If after the first screen you place a second screen with a slit parallel to the first slit, the result will not change. But if the second slit is at right angles to the first, it will block the vibration, so that past the second slit the string will be flat.

The same principle applies in the case of radiation. An electromagnetic wave normally vibrates in random and constantly changing planes, but if it is polarized, it vibrates in a specific plane. The meanings of "partially polarized radiation" and "degree of polarization" should be fairly obvious.

There are materials that polarize light; for example, the well-known Polaroid used in sunglasses. Solar radiation, which is unpolarized, becomes polarized by passing through the Polaroid. When you look through a Polaroid you do not see anything happen, but you perceive that the light is weaker; this is because only 50% of the radiation comes through. Now take a second Polaroid, line it up with the first, and rotate it around the line of sight. You will see the light become gradually fainter and almost disappear; the light is faintest when the second Polaroid is rotated 90°.

This is not the only way that light, and radiation in general, can be polarized. What concerns us is that polarized radiation has well-understood characteristics and that it can be detected. (The detection devices differ, however, from case to case, depending on the wavelength. Polarization is the same phenomenon whether the wave be radio or light, but each of these waves requires its own detection technique.)

Figure 64
Polarization of a wave.
Replacing the screens with
polarizers and the string
with a light beam gives a
model of the polarization of
light.

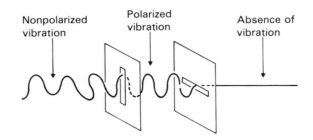

Nonpolarized vibration

Polarized vibration

Absence of vibration

·THE PULSARS

Let us pick up our story where we left off. If you recall, we had reached the conclusion that the intense radio emission from the Crab nebula is synchrotron radiation. This obviously does not fully solve the problem; it just shifts it. Where do the electrons come from? And what produces the magnetic field? Strange as it may seem, the acquisition of knowledge is almost always a shifting of problems. Until we arrive at the prime causes and attain total understanding, it cannot be otherwise. In the process we uncover the countless relations that bind all things into a whole, so that anything that happens has an infinitude of consequences that impinge more or less directly on everything alse. The compart-mentalization of knowledge into branches, sections, chapters, and paragraphs is a useful, even necessary device, but it is totally artificial. It is only due to our inability to describe simultaneously and in their entirety all the dynamic phenomena that occur in the world.

To continue. The development of space technology made it possible to observe celestial objects in the x-ray range of the spectrum; naturally, the Crab nebula was high on the list of celestial objects to be observed. In 1964 a rocket-borne detector revealed that the nebula was also a strong emitter of x rays.

Obviously, 900 years after the supernova explosion there was still something there that supplied the energy necessary to produce such strong radio and x-ray emissions. The solution of the problem came a few years later with the discovery of the pulsars.

The first pulsar was discovered in 1967, by chance, with the new radiotelescope at Cambridge, England. Time and again the use of a new instrument has opened unsuspected avenues of research and brought about results that have had a fundamental impact on scientific thought and hence on our view of the world. I do not think examples are necessary. In this particular case the consequences were not so revolutionary, but it is a fact that the new instrument discovered objects that previously had existed only on paper as theoretical predictions about stellar evolution. Other discoveries followed, but we shall not talk about them because they concern the extragalactic world. This is recent history, and it is too early to tell whether any of these findings will affect everyday life or just add to our store of knowledge (which is no small achievement, to be sure).

The Cambridge instrument was different from previous radio-telescopes in one fundamental respect. No detector has an infinitely fast time of response. It is the same with our sensory organs. Take sight, for example. From the time your eyes spot an obstacle to the time the signal reaches your brain, a fraction of a second elapses. What if the signal requires a response? It takes another fraction of a second for the response signal to reach, say, your feet—about a thirtieth of a second altogether (if you have fast reflexes). If you had been running, this may not be enough time for you to stop short in order to avoid a collision with an obstacle. (In that thirtieth of a second you keep running as if you had seen nothing.) It is the same with instruments. If signals arrive faster than the minimum time a detector needs to register them separately, they appear as one, more or less complex signal. A simple device for recording the radio signals coming into a radiotelescope is the running tape with a marking pen. The pen needs some time to trace a mark on the paper—say, a second. If the incoming signals each last a hundredth of a second, the instrument cannot register them separately. Today, of course, there are much more sophisticated recording techniques, but fifteen, twenty years ago the situation was approximately as I described. By means of special electronic techniques the Cambridge instrument achieved a much faster response time—or, as we say, a much higher resolving power—whereby it could distinguish signals coming at intervals of a thousandth of a second. Furthermore, it had a very low background noise. What is that? If you have carried on a conversation in a crowded room, you know what background noise is. And you know that it can be strong enough to drown out most of your words. In a radio transmission it is the buzzing sound that can interfere with reception. Take a cheap radio and try to hear the pianissimo movements in an orchestral piece. It is no good; they are lost in the background noise. Turning up the volume is no help because the noise becomes louder as well. It takes electronic components to weed out the noise, and then even the weakest signal can be heard. Since the energy coming from celestial objects is very low, we need low-noise instruments with large collecting areas. The Cambridge instrument had all of these features, but clearly fine time resolution was the most important one.

In 1967 this radiotelescope discovered a new type of radio source. Radio sources were nothing new, but this one was very special. Instead of a continuous, more or less regular signal, it emitted an endless series of pulses at incredibly regular intervals, specifically, every 1.33728 sec. This is the kind of finding that would bowl anybody over. The Cambridge scientists got so excited that for a moment they even thought of artificial emitters—extraterrestrial intelligence, in short. The possibility that they could be spurious signals, coming from the earth, that is, or the instrument itself, was quickly discounted after a thorough examination of all the possible sources. Shortly thereafter more pulsating radio sources were discovered. That pointed to too many extraterrestrial civilizations! There had to be a natural explanation for the event. In 1968 one such source was observed close to the center of the Crab nebula, and by 1969 the known sources numbered about 40. Although their periods of pulsation were all very short, they were not all alike, ranging from 0.033 to 3.97 sec. Today these pulsating radio sources, or pulsars, are fairly well understood. Although slow changes in their periods have been observed, these objects pulsate with amazing regularity. The time of arrival of a pulse can be predicted weeks in advance with an error on the order of 10 milliseconds.

Some of these objects have been found to pulsate not only in radio but in other frequencies as well. The pulsar in the Crab nebula, which has the shortest known period (0.033 sec), pulsates also in the visible and the infrared. Figure 65 shows the optical pulsation of the Crab pulsar. Maximum luminosity corresponds to maximum radio emission. It is almost unbelievable to see a star appear and disappear every 0.033 sec! This pulsar was already known as a faint star (magnitude 16), but its optical variability had never been detected. Normal exposure times are too long for such rapid fluctuations. Because of certain spectral features, however, it was suspected of being the relic of the supernova.

It is important to note that such a rapid succession of pulses implies an emitting source of small dimensions. These dimensions could characterize either a limited region of the star or the star itself. The first possibility was discounted because no local phenomenon is known that can last so long with such regularity and be so intense as the one observed. Furthermore, it would mean that a portion of the star would be radiating thousands of times more than the whole star. Consequently, it seems more reasonable to assume that the variability involves the star as a whole. Besides, we already know regular variables whose light variation is not tied to a particular region. What is startling in this case is the extreme shortness of the period. Whatever the explanation may

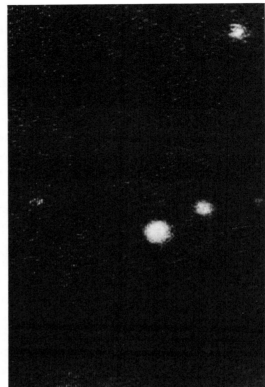

Figure 65
The pulsar in the Crab nebula. The two photographs
were taken between flashes
(left) and at the moment of
a flash (right). As explained
in the text, the time interval
between two consecutive
pulses is 0.033 sec.

be, we must start on the assumption that it is a very small star. Let us see why it has to be so and just how small it must be.

Suppose we receive from a celestial object a pulse that lasts 1 sec. Then the diameter of the object cannot exceed 300,000 km (that is, 1 sec $\times$ c, where c = the velocity of light = 300,000 km/sec). This is the reason. Assume that the pulse is emitted by the entire object all at once, in one instantaneous outburst.[52] Clearly, we receive the pulse from the point of the object that is nearest to us before we have received it from the point that is farthest. Thus since the pulse we receive lasts 1 sec, the farthest point must be 1 light-second away from the point nearest. Should the sun suddenly be turned off, we would not see all of it disappear in the same instant because the light from the edge of its disk takes longer to reach us than the light from the center of its disk—2.5 seconds longer, since the sun's radius is 750,000 km. The center of the disk would disappear first, and then we would see a dark round spot expand around it and cover the whole sun in the space of 2.5 seconds. Conversely, if the entire sun emitted a light pulse, the pulse could not last less than 2.5 seconds. Naturally, a region of the sun could emit very short pulses, but in any case the dimensions of the region could not exceed the length traveled by light in the time the pulse lasts. Consequently, the time of duration of a pulse gives an upper limit to the size of the emitting object.

The pulses emitted by the pulsars last on the order of 20 milliseconds (at intervals ranging from 0.033 to 3.97 seconds, depending on the star). On the assumption that a pulsar is a sphere, its radius cannot exceed the distance that radiation covers in 20 milliseconds, which is 6,000 km. Since in reality the pulse is *not* instantaneous, the star's radius must be smaller than that. Theoretical considerations actually show that a pulsar must be much smaller than the earth, with a radius on the order of tens of kilometers. If you recall, the range in stellar masses is not very wide, and for now we can safely assume that the mass of a pulsar is roughly of the order of one solar mass. If so, we are confronted with a very peculiar object. A star of one solar mass and a radius of some 10 km has an enormous, inconceivable density. What then is a pulsar? The current interpretation is discussed in part III as part of a panoramic view of stellar evolution.

·PULSARS AND SUPERNOVAE

All available evidence points to the conclusion that a supernova is an exploding star. Furthermore, the energy emitted in the course of the explosion is an appreciable fraction of the energy a normal star emits in its entire life. Therefore, in spite of the name, which suggests a particularly intense nova, a supernova is an entirely different phenomenon. The pulsar in the center of the Crab nebula is a new object whose behavior is not yet fully understood. It is an exceedingly small star (see figure 66) that emits radio, optical, and x-ray pulses; x-ray emission has also been observed from the surrounding nebula.

All in all, the nature of the supernovae is still somewhat obscure. Many supernova remnants do not contain a pulsar, whereas many pulsars appear to be surrounded by supernova remnants. Some remnants emit x rays (the Cygnus nebula, for example; see figure 59), but many do not. Most x-ray sources are not supernova remnants; finally, not all supernovae seem to end up as pulsars. Clearly, the problem of the supernovae cannot be completely solved until enough data are gathered to put the matter on a good statistical basis. Meanwhile, theoretical astronomers will argue, propose interpretations, and make models. One day, perhaps, new observations will yield the information needed to choose among the various interpretations, and whoever has been fortunate enough (so to speak) to propose the model that best fits the facts will be remembered as the one so brilliant as to see it all. But it is obvious that nobody can really predict how things will turn out and that making models "before the fact" is almost always like playing the numbers game. Not that it is a useless exercise— not by any means. Indeed, these games have often inspired experiments that eventually opened new avenues of research and even altered the course of science.

Figure 66
Comparison of the size of the earth with the size of a pulsar (indicated by arrow).

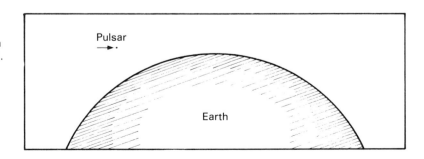

176 • THE GALAXY

·STARS WITH AN EXTENDED ATMOSPHERE; THE PLANETARY NEBULAE

Let us continue with our survey of galactic objects. There is one more type of star worth mentioning, namely, the star with an extended atmosphere. Stars of this type, although surrounded by very large and expanding gas envelopes, do not display the characteristics of explosive events. In some cases the envelope can be observed directly with the telescope, but these stars are most often recognized from the spectrum, which shows the bright lines typical of thin gaseous masses; the underlying spectrum is produced by the stellar body. Sometimes, instead, the presence of the shell is revealed by very narrow absorption lines. Several red giants and a few thousand stars of spectral class B have an extended atmosphere.

In the case of the so-called shell stars, spectroscopic observations reveal a marked broadening of the lines due to the Doppler effect. This fact suggests that a star of this type rotates very rapidly about its axis. Indeed, such stars can have surface velocities on the order of 200 km/sec, compared with 2 km/sec for the sun. At the star's equator, which is the locus of points on its surface farthest from its rotation axis, the linear velocity is highest; in fact, it is approximately the velocity a satellite must have in order to revolve around the star without falling into it. According to the current interpretation a small perturbation causes the star to release a gas shell, and as this shell expands, others form, so that the star is eventually surrounded by a series of expanding shells.

The planetary nebulae are the best-known group of stars with an extended atmosphere; typical examples are the famous ring nebula in Lyra (figure 67) and NGC (New General Catalog) 7293 (figure 68). These objects were called planetary nebulae because when first observed (with small telescopes) they looked like small disks similar in appearance to the planets. The name has been retained although these objects have nothing to do with planets. In reality, as modern telescopes show, a planetary nebula consists of a star surrounded by a slowly expanding (15 to 50 km/sec) nebula. Although spherical, the envelope has an annular look. To understand why, think of a soap bubble, which is transparent at the center (where the line of sight crosses a thin film) and more opaque at the edges (where the line of sight has to cross a greater thickness).

Figure 67
The planetary nebula in
Lyra.

Figure 68
The planetary nebula NGC
7293.

Most of the known planetary nebulae have been discovered by spectroscopic observations because the telescope can only distinguish the nebula from the star when the object is near. The majority are point sources like the other stars. Less than a thousand are currently known, but it is obvious that there must be many more—perhaps a few tens of thousands—because the most distant ones elude our observations. Nevertheless, they are rare; a few tens of thousands is a small number compared to the 100 to 200 billion stars in our galaxy. This means that planetary nebulae are generally very distant objects. Since they are rare, and since there is no reason why they should be concentrated in any particular place, it is reasonable to expect that most of them will be far from any random observation post, such as the earth. As a result, trigonometric parallaxes can only be measured for a few of them.

Furthermore, these peculiar objects do not fall in the classical domains of the H-R diagram, so that their spectroscopic parallaxes are also difficult to obtain. Consequently, their distances have to be determined by statistical methods and are not known with good accuracy. Distances and angular measurements of diameters, when possible, give the dimensions of the nebulae, which range from 0.5 to 1 light-year. From the size and the expansion velocity of a nebula and also from the assumption that this velocity has remained constant over time, we can work backward to find the age of the nebula, that is, the time that the gas began to be ejected from the star. A typical age is 20,000 years, and most known planetaries have formed in the last 50,000 years. This makes sense. After 100,000 years the nebula has expanded so much and has become so tenuous that it can no longer be seen, so that a star that was once a planetary nebula no longer looks like one. This explains why planetary nebulae are so rare. By cosmic standards they are short-lived events, which probably involve only the stars' surface layers.

The central body in a planetary nebula is a hot blue star that emits a very intense spectrum in the ultraviolet. On the other hand, the radiation from the nebula corresponds to a much lower temperature than that of the star and therefore must be produced by fluorescence. For those of my readers who are not familiar with this phenomenon, I shall try to explain it by what was said about the origin of radiation and the energy levels of the atom.

Suppose we observe a hydrogen atom. (Any other atom would do, though.) Suppose, further, that our atom has been excited by the capture of an energetic photon and has jumped from the first (ground) level to the fourth. It can return to the first level either in one jump, emitting a photon of the same wavelength as the captured photon, or by intermediate stages. Let us say that it jumps from the fourth level to the second and then from the second level to the first. In the first jump ($4 \rightarrow 2$) it emits a photon of energy $E_{4 \rightarrow 2} = h\nu_{4 \rightarrow 2}$ (equivalent to the difference in the excitation energies of the two levels); in the second jump it emits a photon of energy $E_{2 \rightarrow 1} = h\nu_{2 \rightarrow 1}$. We know that $E_{4 \rightarrow 1} = E_{4 \rightarrow 2} + E_{2 \rightarrow 1}$, that is, $h\nu_{4 \rightarrow 1} = h\nu_{4 \rightarrow 2} + h\nu_{2 \rightarrow 1}$. This means that a photon of energy $h\nu_{1 \rightarrow 4}$ ($= h\nu_{4 \rightarrow 1}$) has been absorbed and that two photons are emitted whose energies are, respectively, $h\nu_{4 \rightarrow 2}$ and $h\nu_{2 \rightarrow 1}$. Thus the absorbed photon has a higher frequency than either of the emitted photons, which is the same as saying that the absorbed photon has a shorter wavelength than either of the emitted photons. Considering now the nebulosity as a whole, the hydrogen gas absorbs radiation of short wavelength and converts it to radiation of longer wavelength.

This is the process of fluorescent emission at work in a planetary nebula. The total amount of visible light radiated by the nebula is much higher than that emitted by the central star. Since the star is the only source of energy of the nebulosity, it must be concluded that the star is so hot that most of its energy is emitted as invisible ultraviolet radiation, which is then absorbed by the nebular atoms and reemitted as visible light. (In actuality, at least for thousands of years, only the inner part of the nebula emits visible light because only here is the gas dense enough to absorb the radiation from the star; the outer layers are cold and transparent.)

The central star is indeed very hot, as could be deduced from its blue color. If we work backward from the emission of the nebula and explain it on the basis of fluorescence, we find that the surface temperature of the star must be at least 20,000°K. Some of these stars have been found to have temperatures in excess of 100,000°K, which makes them the hottest objects in our galaxy. Since, all in all, these stars are not very luminous, they must be quite small. A few are actually comparable in size to white dwarfs. We shall come back to them later.

·A LOOK AT THE GALAXY

Before we complete our survey of galactic objects, we should take a panoramic look at our galaxy. In this book I am not going to

discuss our galaxy per se. To do so we would have to look at the sky from a different perspective and switch fom a microcosmos to a macrocosmos in which individual stars are as unimportant as single cells in the study of the human body. A further step would be to view our galaxy, in turn, as a microscopic element of that entity made up of all the galaxies and intergalactic space we call the universe. Let us just talk a little about our galaxy, enough for you to visualize it and place in it the various objects we have been discussing.

The Milky Way is basically an aggregate of about 100 to 200 billion stars. Similar stellar systems are scattered by the millions through space, separated by distances of millions of light-years. By looking at the sky through a telescope, Galileo was the first man to realize that it was far more complicated than it seemed. Early in 1609 Galileo heard about an optical instrument built in Holland that made distant things seem close by, and grasping its potential, he reinvented it. Although this was no small achievement in itself, Galileo's greatness lies in the fact that he chose to use it for astronomical observations (figure 69). In 1610 he described what he had seen by pointing the telescope at the Milky Way; the luminous belt stretching across the sky was in fact composed of distant stars, myriads of them.

Today we know what this band of stars is. Suppose you are inside a stellar system shaped like a thin disk. If you look in a direction perpendicular to the plane of the disk you see few stars. But they become more and more numerous the more your look shifts to a direction parallel to the disk's plane. If the space density of the stars is sufficiently great and/or the dimensions of the disk sufficiently large, by an effect of perspective you see a concentration, almost a continuity of stars. The first person to have the right idea about the Milky Way and the shape of our galaxy was T. Wright in 1750. Kant proposed a similar hypothesis in 1780, and in 1785 W. Herschel proved its validity by means of star counts performed for 683 selected regions of the sky. He found that some regions contained as few as 1 star, and others as many as 600. There were two ways of explaining these findings: Either the average density of stars per unit volume of space is widely different in different directions, though not random, or the extent of the space occupied by stars is not the same in all directions. He chose the latter alternative because it was simpler. Everything could be explained if one assumed that the stars formed a system in the shape of a disk, as illustrated in figure 70. If the observer

Figure 69
Two of Galileo's telescopes. Preserved in a frame is one of the lenses he used.

Figure 70
Early view of our galaxy.

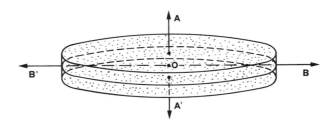

is at O, the line of sight in the direction OA (perpendicular to the disk's plane) might not meet any star, while this is practically impossible in the direction OB (parallel to the disk's plane). Indeed, since for the observer all the stars are at an infinity, by revolving the line of sight from the direction OB to the direction OB', thereby scanning a circle parallel to the disk's plane, the observer sees a band of light on the celestial sphere, a Milky Way, that is the sum total of all the background stars that the eye places on the circle containing B and B'. The disk Herschel envisaged had the sun at the center; obviously something had remained of the old geocentrism, although now it had turned to heliocentrism.

Our galaxy is not shaped exactly the way Herschel had imagined, nor is the sun in its center. Since we are inside it, it is no easy matter to determine its structure, and our current ideas of it are based on observations of external galaxies as well as on direct observations. If we could go outside it (I almost said "into the open"), we would see that our galaxy is shaped like a spiral with a central bulge 3,000 parsecs thick and spiral arms wrapped around it. The sun is located in a spiral arm approximately two-thirds of the way (10,000 parsecs) from the center to the circumference. Our galaxy measures 30,000 parsecs across, and if you are not impressed by this number, think of it as 100,000 light-years, or 10^{18} km, or a billion billion km. That should leave you gasping! In other words, it takes light 100,000 years to travel from one end of it to the other. The shape of our galaxy is essentially due to its rotation about its polar axis. It takes the sun about 200×10^6 years to complete a full revolution about it at a speed of about 250 km/sec. Unlike a wheel or a spinning top, however, our galaxy does not rotate as a rigid body. Hence the linear velocities of the stars are not simply proportional to their distances from the center.

The stars in our galaxy are sometimes grouped into systems known as clusters,[53] which are distinguished in open, or galactic, clusters and globular clusters. The latter are actually located outside the disk in a spherical region called the galactic halo or corona, whose equatorial radius is greater than that of our galaxy. The halo also contains some gas and scattered stars, but the concentration of objects decreases as the distance from the disk increases.

Beside stars, our galaxy contains gas and dust, which together make up what is known as the interstellar medium. This interstellar matter is extremely patchy. To begin with, the gas and dust are concentrated in the spiral arms. Furthermore, here and there, but particularly in the spiral arms, the material collects in vast clouds, called nebulae, that can be more than 100 light years across. There are both dark and bright nebulae.

Figure 71 summarizes what I have said about our galaxy. Figure 72, which is a photograph of the Andromeda galaxy, clearly shows that galactic matter—in any form—is not uniformly distributed. It is easy to see that the density of matter is much lower between the spiral arms than in the arms themselves.

After this quick look at our galaxy, we shall stop and investigate two members of the galactic family we have just met—the interstellar medium and star clusters.

·INTERSTELLAR GAS AND DUST

Gas and dust make up the interstellar medium that fills the space between the stars. "Fill" may be the wrong word in this case because interstellar space is practically empty. One cubic centimeter of air at sea level contains 3×10^{19} molecules. A cubic centimeter of interstellar gas contains on the average a mass equivalent to a hydrogen atom. To understand how large a number 3×10^{19} is, suppose you want to take out all the air molecules; at the rate of 100,000/sec, it would take you 10 million years; at the rate of 100 million/sec, it would take you 10,000 years. Compared with 10^{19}, 1 particle/cm^3 is practically nothing. Our galaxy is so vast, however, that the total mass of the interstellar medium is something like 4 billion suns. Collectively, all the stars have a mass of about 100 to 200 billion suns. Hence the interstellar medium is a significant component of our galaxy. Sometimes this material collects in clouds whose densities can be 50, 200, or 300 times greater (see plates 7–11). Even so, we are dealing with matter so rarefied that no vacuum pump on earth could possibly produce it. The temperature of the interstellar gas is about 50°K (about −200°C).

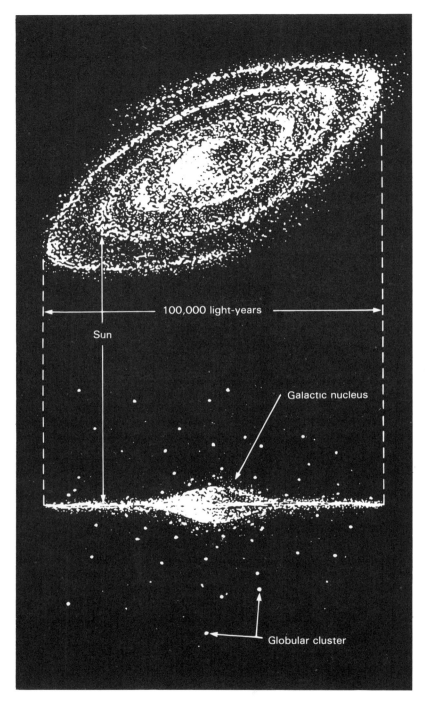

100,000 light-years

Sun

Galactic nucleus

Globular cluster

Figure 71
Schematic illustration of our galaxy, in views from above (top) and the side (bottom).

Figure 72
The Andromeda galaxy, which in all likelihood is very similar to our own. It is at a distance of about 2×10^6 light-years from us and is approaching at the speed of 300 km/sec. Andromeda has a mass of about 3×10^{11} solar masses.

This prompts me to remark on a peculiar aspect of astrophysics that is of fundamental importance in the study of matter. Astrophysics deals with extremes of density and temperature and, in general, with extreme conditions for most of the parameters that describe the states of matter. The stars, the interstellar medium, and the galaxies are unique laboratories in which phenomena occur that could never be produced here on earth in our laboratories. Thus the sky offers the opportunity to study matter under conditions impossible to duplicate, and so to acquire the kind of knowledge that is the ultimate test of physical and chemical theories. On a different level, I might add, astrophysics also provides the grounds for testing the meaning and validity of philosophical ideas, religious beliefs, and other aspects of human thought.

The interstellar gas is similar to the main-sequence stars in composition—1 helium atom for every 10 hydrogen atoms plus traces of other elements—and in weight—3 parts hydrogen to 1 part helium plus a small percentage of the other elements.

As just noted, when the interstellar matter is denser, clouds, or nebulae, form. These can be of different types. There are vast clouds, many parsecs across, which are composed in good part of dust. As figures 73 and 74 show, dust clouds can completely obscure the light from the stars behind. The farther away these dark dust clouds are, the more difficult it is to observe them with optical instruments; this is also because as a cloud's distance from us increases, so does the number of stars between it and us. Sometimes there are so many foreground stars that they cover the field of view almost entirely. To have an idea of what I mean, look at figure 75. In addition to all this, distant clouds are also obscured by the diffused interstellar medium.

A particular type of dark cloud is the so-called globule, which can be observed on the light background of bright nebulae (which will be discussed shortly) or rich stellar fields. Dark clouds of this type were named globules because of their round or oval shape and small angular size (a few seconds of arc; up to an upper-limit distance of 1,000 parsecs, this corresponds to a size ranging from a few thousand to 100,000 astronomical units). The globules are very dense compared with the normal formations of dark clouds, since they dim the light behind them by 5 magnitudes or more.

Absorption of the radiation coming from the stars is not caused by dark clouds alone. Interstellar dust, which is diffused everywhere, causes a general dimming of starlight, and the dimming is obviously more pronounced for distant stars. The galactic center

Figure 73
The Horsehead nebula in
Orion. It is a dark cloud visible on the bright back-
ground of a vast HII region.

Figure 74
Region of the Milky Way in
the constellations of Serpens and Aquila. Note the
large dark clouds that blot
out the stars behind them.
The stars that dot the dark
clouds are foreground stars.
The light regions that look
like bright nebulae are
actually clouds of stars.

Figure 75
Star field in Sagittarius,
showing at center left the
Lagoon nebula and at lower
left the Trifid nebula.

is so obscured by interstellar dust that it cannot be observed in the visible region of the spectrum. If there were no dust, our night sky would be much brighter.

One important consequence of interstellar absorption is that the stars appear less bright than they actually are. If the apparent magnitude we assign to a star is higher than it should be, its intrinsic brightness will be underestimated. Conversely, if the absolute magnitude is known (from the H-R diagram, for example), the distance will be overestimated. If you recall (see part I on steller magnitudes and note 45),

$$M = m + 5 + 5 \log p;$$

thus if M is given and m is greater than it should be, $5 \log p$ turns out to be smaller than it should be, and the distance, which is the reciprocal of p (parallax), is greater. Hence the importance of estimating the extent of interstellar absorption.

One way of making this estimation involves a painstaking job of stellar counts. The job, in essence, means selecting certain regions of the sky (away from the galactic plane) and counting: first the number of stars up to a given magnitude; and then increasingly fainter stars. On the assumption that the stars are more or less uniformly distributed throughout space and that they all have the same intrinsic brightness, one expects to find an increase in number as fainter and fainter stars are taken into account. This increase is observed, but it is not as fast as expected. This is because of interstellar absorption. In this manner it is possible to estimate average absorption in different directions.

A better way to determine interstellar absorption is based on the so-called interstellar reddening. You may have noticed that a street light becomes redder, the farther you move from it. Here is a better example: if you observe street lights from a high place, you will notice that the farther away they are, the redder they become. This is due to the scattering of light by atmospheric particles. We have already discussed scattering to explain the blueness of the sky and the redness of the setting sun. Blue rays are deflected more than red rays. With the blue rays partly or wholly removed, the sun, or the street lamp, looks redder than it would be in the absence of the atmosphere. Similarly, interstellar dust has the ability to redden starlight.

Suppose we obtain the spectrum of a distant star. From the absorption lines we can determine the spectral type, which tells us the temperature of the star and hence its true color. This, in turn, gives the color index that the star should have. By comparing this color index with the observed color index,[54] the degree of reddening can be estimated with good accuracy. Thus we can estimate how much dust the light of the star has encountered on its way to the earth and by how much the star has been dimmed. Instead of using a spectrum, which is difficult to obtain when the star is too far and too faint, we can use a photometer to make observations at three different wavelengths. Then, by making the appropriate corrections in our formula to take into account space absorption, we can find the actual distance of the star.

Although we usually say that dust clouds absorb starlight, the prevailing mechanism is in fact scattering. We speak of absorption because for the observer the effect is the same—a dimming of the light emitted by a star. But if there is scattering, it should produce the same effect that we observe in our atmosphere, namely, the cloud should be illuminated. The fact is that our atmosphere consists of gas, and a gaseous mass is at least 100,000 times less opaque than an equal mass of dust. Furthermore, the particles in a gas—atoms or molecules—are much smaller than dust grains. Consequently, the behavior of a gaseous mass is basically different from that of an equal mass of dust. Gas reddens light much more than dust. (That is, gas scatters blue light much more than dust.[55]) But the dust is much more efficient at blocking out the light. So much so that if we wanted to explain the dimming of starlight by the presence of gas, we would have to invoke an enormous amount of it—enough to have gravitational effects on the motions of stars embedded in a cloud. No such effect has ever been observed. We also know by everyday experience that dust grains, and large particles in general, are much more efficient at dimming sunlight; smoke and soot as well as fog and clouds (water vapor condensed into drops) obscure the sun much more than the entire atmosphere, which is a very large mass of gas. To conclude, it is dust rather than gas that is responsible for obscuring the stars.

As I was saying, scattering of the light that crosses a cloud should cause the cloud to light up. Generally, however, the level of light is so low that the presence of a cloud can only be deduced from the dimming of the stars behind it. But when a cloud chances to be near one or more bright stars, it lights up to the point that it can be seen and photographed. In this case, the cloud is known as a reflection nebulosity. Typical examples are the nebulosities surrounding the Pleiades (see figure 76). Since inside such a cloud

Figure 76
The Pleiades and associated
nebulosities.

blue light is scattered more than red light, a nearby star looks a little redder than it really is (having lost more blue light than red), and the cloud looks a little bluer than the star. Only a very red star (a cool star) can be surrounded by a reddish nebulosity, while only very hot white or blue stars can have blue nebulosities.

What is the chemical composition of the dust grains? The question has not been fully settled yet. Since we cannot get a sample of the interstellar medium and analyze it, research is carried out on the basis of theoretical considerations and laboratory experiments. Some information has been recently obtained from observations in the ultraviolet performed by OAO (Orbiting Astronomical Observatory) between 1968 and 1972. What we are looking for is the combination of elements and compounds that best explains the observed dimming both in quality and quantity. But the possible combinations are almost endless. Furhermore, it is not enough to find a plausible combination; it must also be in accordance with current theories on the formation and dispersal of elements in our galaxy and with our observations of the stars. Finally, the proposed combination must be stable; if the material we envisage is not going to last in the physical conditions of interstellar space (if it is likely to vaporize, for example), obviously it cannot be a satisfactory solution. Graphite appears to be a good candidate, but it cannot be the only one. Certain hydrogen compounds may also be present. I think I have said enough to indicate the complexity of the problem.

Let us turn now to the other component of the interstellar medium. There are great quantities of gas in our galaxy, but as we said, it is so transparent that it does not cause any appreciable dimming of starlight. Its existence is revealed, however, by spectroscopic observations.

In addition to the dark Fraunhofer lines produced by the sun's own atmosphere, the solar spectrum displays absorption lines due to components of the earth's atmosphere. This is because on its way to the spectroscope, sunlight must cross our atmosphere, so that solar photons of certain wavelengths may be absorbed or scattered by atmospheric particles. At these wavelengths, therefore, we observe a weakening of sunlight, that is, absorption lines. It is fairly easy to distinguish the lines of solar origin from those of terrestrial origin. Observing the sun's spectrum in the course of a day serves the purpose. When the sun is low on the horizon, its light has to cross a thicker layer of atmosphere than when it is high; consequently, the lines of terrestrial origin become more intense (since there is more absorption) during the afternoon. Thus in order to identify the lines produced by atmospheric absorption, all we have to do is to observe which lines become stronger as the sun sets.

Analogously, interstellar gas absorbs starlight and imprints absorption lines on stellar spectra. The gas is very tenuous, but starlight has a long way to travel. Sooner or later the photons that interstellar gas can absorb will be absorbed, and at these wavelengths there will be a shortage of light; hence absorption lines appear on stellar spectra. The observation of interstellar lines is confined chiefly to the more distant stars because an exceedingly long column of gas is needed to produce absorption lines strong enough to be visible. Interstellar lines are best seen in the spectra of hot stars, which are relatively simple. The spectra of cooler stars are very complicated and their lines are so numerous and so crowded together (especially if the spectra are obtained with low-dispersion spectrographs) that it is very hard to distinguish among them the interstellar lines, which, moreover, are generally weak. The lines produced by interstellar gas can be recognized by their Doppler shift, which is different from that of the star's own lines because gas clouds do not move at the same speed as the stars. Of course, like all other spectral lines, interstellar lines can give us an indication of the chemical composition of the gas that produces them, as well as information concerning its temperature, density, and motion.

As we shall see later on, the interstellar gas can also be detected from its radio-frequency emission, particularly at the wavelength of 21 cm. Finally, when a fairly dense gas cloud is in the vicinity of one or more bright stars, it can become clearly visible. This happens because the gas absorbs the radiation from the star and reemits it essentially by the same mechanism (fluorescence) as that involving planetary nebulae. Such bright clouds are known as emission nebulae.

The interstellar gas consists of the common elements, but hydrogen is by far the most abundant, followed by helium. Together they make up 95 to 99% of its mass. Three-quarters of the interstellar gas is hydrogen and therefore hydrogen is what we find in emission nebulae.

Due to the low temperature of interstellar space, the hydrogen found in it is generally in the neutral state and at the fundamental energy level. Under these conditions it is invisible, that is, cannot be seen with optical instruments, though it can be detected by radio telescopes.

Let us see what happens when this neutral hydrogen is near a very hot star that emits great quantities of ultraviolet radiation. An ultraviolet photon of wavelength shorter than 912 Å has enough energy to tear the electron away from the lowest energy level of a neutral hydrogen atom. When a neutral hydrogen atom absorbs such a photon, it becomes ionized; that is, it splits into a proton and a free electron. In themselves, these two particles do not emit radiation. Given the relatively high density of interstellar clouds,[56] however, there is a good chance that they will recombine with electrons and protons produced by other processes of photoionization to reconstitute hydrogen. If on recombination the electron lands in one of the higher orbits, it will cascade to lower orbits, emitting photons as it jumps. Thus the gas fluoresces and becomes luminous. The electrons can make all the possible jumps, and therefore the gas can emit all the possible hydrogen lines; but only the lines of the Balmer series (see figure 36) fall in the visible range of the spectrum. Ultraviolet emission can only be observed outside the earth's atmosphere.

The reconstituted hydrogen atoms do not last very long because they soon become ionized again by the absorption of ultraviolet radiation. At any given time, therefore, the nebulosity is essentially composed of protons and electrons, or, as we say, ionized hydrogen. Since the symbol of ionized hydrogen is HII (HI is neutral hydrogen), these regions of emitting gas are called HII regions. I would like to call your attention to the fact that although HII regions consist of ionized hydrogen, the emission from these regions is due not to the ionized atoms, as often stated, but to electron jumps in the neutral atoms.

Two of the most famous emission nebulae are the Orion nebula (figure 77 and plate 12) and the North America nebula in Cygnus (figure 78).

It must be clear by now that near a vast emission nebula there must be very hot stars of spectral class O. Since the luminosity of the nebula depends on the intensity of the ionizing radiation, it follows that B0 and B1 stars have smaller nebulae; starting from spectral class B2, emission nebulae, if any, are smaller yet. If there is a visible nebulosity around a cooler star, it is because the starlight falling upon the cloud is scattered by dust particles; this is a reflection nebula, and the spectral composition of its light is about the same as that of the illuminating star.

Figure 77
The great nebula in Orion,
which is both a reflection
nebula and an emission nebula (HII region). It is (relatively) very dense—300
atoms/cm³ halfway be-
tween the center and the
edge. The nebula is 8 parsecs across and is at a distance of 500 parsecs.

Figure 78
The North America nebula,
which extends 40 light-
years and is at a distance of
650 light-years. The dark
region that defines the "At-
lantic coast" and the "Gulf
of Mexico" is a dark nebula.
Seen on the right is the Peli-
can nebula.

The extent of an HII region depends on the luminosity of the star, or stars, in the ultraviolet, but also on the density of the gas. If the gas is very dense, the ultraviolet radiation is absorbed within a relatively small space. But if the cloud is tenuous and the star is very hot and luminous, the HII region (emission nebula) can be quite vast. For example, if the density of the gas is of 1 hydrogen atom/cm^3, which is the average density of the interstellar gas, a main-sequence star of type O6 can ionize a region 100 parsecs across. In the same conditions, on the other hand, a B0 star would produce an HII region 40 parsecs across; and an A0 star, an HII region 1 parsec across. Clearly, the temperature and luminosity of the star are the determining factors.

The interstellar gas can also be detected by its radio-frequency emission. In the wavelength range from 1 to 30 m, we receive radiation of galactic origin that is emitted mostly by the interstellar gas rather than by the stars. Discrete (localized) radio sources exist, of course. A strong radio source we have already mentioned is the Crab nebula; others are Cassiopea A and the nebula in Orion. The sun, too, is a radio source, and a pretty strong one owing to its proximity. But even if all the stars were radio sources as strong as the sun, their combined radio-frequency emission would be but a billionth of the total energy we receive in this portion of the electromagnetic spectrum.

There are various ways in which radio waves are produced. One mechanism we discussed with regard to the Crab nebula is acceleration of the electrons by a magnetic field (synchrotron radiation). Free-free transitions of the electrons in a proton field in an HII region can also originate radio waves. If you recall, a free electron passing close to a proton may be affected by its attraction and change course. If in the process the electron loses energy, a quantum of radiation is emitted whose frequency, or wavelength, depends on the lost energy according to the formula $E = h\nu$. Encounters of this type involve small energies—and hence emissions at radio-frequency wavelengths. Since there are no restrictions on the energy levels that free electrons can occupy, all transitions are possible, and therefore the emitted spectrum is continuous. Radio-frequency emission in fact occurs in the wavelength range from 1 to 30 m.

More important to the study of interstellar gas is the radio-frequency radiation of 21-cm wavelength emitted by neutral hydrogen. Let us see by what mechanism this emission originates. We already know the energy-level diagram for hydrogen (see figure 36). A transition that produces an emission at the 21-cm wavelength, that is, a quantum of energy of fairly long wavelength, must involve a very small jump in energy. Figure 36 illustrates that small jumps in energy can only take place when the energy levels are close together. But this type of transition cannot be the answer. In the first place, if a jump between the right levels were to occur, jumps should also occur between nearby levels because it cannot happen that one level is particularly populated while the nearby levels are not. Consequently, instead of a neat line at 21 cm we should observe a band of emission in a small range of wavelengths around 21 cm. But we do not. Furthermore, the interstellar hydrogen is generally at very low temperatures; except in HII regions, therefore, it is neutral and at the fundamental energy level, which is the lowest excitation level (HI regions).

At this point I have to confess that when I discussed the possible energy levels of the hydrogen atom, I left some things out—in particular, the so-called hyperfine structure, which is responsible for hyperfine transitions. An electron spins like a top as it revolves about the nucleus, and it turns out that it makes a difference for the total energy of the atom which way the electron is spinning. The proton (the nucleus in this case) spins too. If both the electron and the proton spin in the same direction, the atom has a higher energy than if they spin in opposite directions. In the hydrogen atom (at the fundamental level) the spins are normally opposite; this is the configuration of lower energy, and hence is more stable. But if the atom absorbs the right amount of energy, the electron may flip and spin the other way. Although it is still at the fundamental level, the atom is now excited, having moved to the upper level of the hyperfine structure (of the fundamental state). As with the other levels, the excited condition has a finite duration; the electron spontaneously reverses its spin, the atom jumps back to the lower level, and in the transition a quantum of radiation is emitted whose energy is equal to the difference in energy between the two levels. Recalling once again that $E = h\nu = hc/\lambda$ (where E is the energy, h Planck's constant, c the speed of light, and λ the wavelength of the emitted radiation), if the energy difference between the two levels is very small, then the wavelength λ of the quantum is very large; in the special case, $\lambda = 21$ cm. This event is very rare. The hydrogen atoms in the HI regions may jump to the upper level of the hyperfine structure by colliding with other atoms, but the interstellar gas is so tenuous that it takes millions of years for atoms to collide with one another.

There are so many atoms in interstellar space, however, that at any given instant the number of excited atoms is very large. There is one additional difficulty. The transition from the upper to the lower level of the hyperfine structure is a *forbidden* transition, so called because the probability of its occurring is exceedingly small and hence never observed in the laboratory. A hydrogen atom excited from the ground level to, say, the third will spontaneously return to the ground level, either directly or in two jumps, in 10^{-7} to 10^{-8} sec. On the other hand, an atom excited to the upper level of the hyperfine structure will remain there on the average for 10 million years before returning to the lower level; this transition is strongly forbidden. But time is plentiful in our galaxy, and the number of atoms beyond count. And even though an excited atom can lose its stored energy by colliding with another atom, the 21-cm radiation turns out to be intense enough to be observed, as H. C. Van de Hulst had predicted in 1944.

The 21-cm radiation is a very powerful tool for studying both interstellar gas and the galactic structure. By scanning the sky with radiotelescopes tuned to this particular frequency, we can study the distribution of neutral hydrogen in our galaxy because the intensity of the line tells us the number of atoms in the line of sight. Radio maps of our galaxy show that the hydrogen is distributed within a layer about 100 parsecs thick that extends along the equatorial plane. By the intensity of the signal, it has been calculated that the mass of the gas is approximately 2% of the mass of the entire galaxy. From the Doppler shift of the 21-cm line, we can study the motions of the gas; one important result of these observations is that our galaxy is indeed a spiral. More about this shortly.

·STAR CLUSTERS AND ASSOCIATIONS

And now let us take a quick look at star clusters; as the name implies, these are aggregates of stars. There are two types of clusters—the open, or galactic, clusters and the globular clusters. They differ in size, distribution in space, numbers of stars, and, most important, the kinds of stars they contain.

Galactic clusters are found mostly near the galactic plane, and their dimensions generally do not exceed 10 parsecs. The number of stars in these clusters ranges from a handful to several hundreds. Sometimes the stars are so thinly spread that the cluster can hardly be recognized against the background. The fact that it is a cluster is often revealed by the motions of the components (which travel in the same direction and with the same velocity) or by an H-R diagram constructed for these stars alone. But what makes discovery of these clusters most difficult is interstellar absorption, which is so strong along the galactic plane that at a certain point observations become practically impossible. About 1,000 galactic clusters are known, but the fact that they are all located within 2,000 parsecs of the sun says that there must be many more in our galaxy—certainly several thousands. Typical examples of galactic clusters are the famous Pleiades, 400 light-years away, which consist of more than 100 stars, six of them visible to the unaided eye (figure 76); the Praesepe cluster, 500 light-years away; and M 67 in Cancer, 2,500 light-years away.

The globular clusters are found all around the disk, in the galactic halo. This makes them easier to observe, except of course when they are near the central bulge of our galaxy, where observations are hampered by interstellar absorption and the great number of stars in that part of the sky (see figure 79). As I mentioned earlier, they are regarded as galactic objects because they are gravitationally bound to our galaxy. About 125 are known, and since they are distributed at random, there cannot be many more. Globular clusters range in size from 50 to 100 parsecs and contain thousands of stars; the richest may have as many as a million. M 13 in the constellation Hercules (figure 80) and M 3 (figure 81) are typical examples.

Unlike the galactic cluster, whose components are scattered far and wide, the globular cluster is spherical in shape (hence the term globular) and its stars are tightly bunched together, with increasing concentration toward the center. Figures 80 and 81 show the central region of such a cluster as one bright body because the stellar images merge there. In reality, of course, the stars in the central region are far apart, so much so that if an outside star were to move into it, nothing untoward would happen; most probably, the intruding star would simply go in one way and come out the other. As always, you must keep in mind the scale of things. The stars are very small compared with the distances that separate them, and even in the center of a cluster, a planet could happily revolve about a star without worrying about possible collisions. The night sky of such a planet would be quite a sight, with so many stars shining like the full moon on earth!

A cluster—galactic or globular—is held together by gravitational attraction, although its component stars tend to go their separate

Figure 79
Region in Sagittarius near
the galactic center. Two
globular clusters, NGC 6522
and NGC 6528, can be seen
through the stars.

Figure 80
The globular cluster M 13.

Figure 81
The globular cluster M 3.

ways by the law of inertia. In the long run, however, a cluster is destined to lose its stars. It is only a question of time, but as I said before, time is plentiful. The stars have a good chance of coming close enough to perturb one another. Such close encounters may cause one star to slow down and another to speed up. Once in a long while, a star will speed up enough to overcome the gravitational attraction of the system and escape. Little by little, the cluster becomes poorer. In addition, the system is subject to disrupting forces originating from masses that are concentrated here and there in the galaxy and especially in the spiral arms. As a result, a cluster cannot last forever; sooner or later it will fall apart.

Galactic clusters tend to break up much faster than globular clusters. Apart from everything else, they contain fewer stars, and the gravitational attraction that holds them together is not as strong. A galactic cluster may be scattered in 500 million years, whereas it takes at least 100 billion years for a rich globular cluster to fall apart. Since our galaxy is believed to have formed 5 to 10 billion years ago, we have to assume that the galactic clusters we observe today formed when our galaxy was already in existence and that the poorer clusters formed before the richer ones. It seems reasonable to conclude that since the time the galaxy came into being, galactic clusters have formed and disappeared while others were forming, and so on. On the other hand, the globular clusters break up so slowly that they may have existed as we see them, though with many more stars, since the formation of the galaxy.

There is a third type of star aggregate known as an association. In such a system the mutual gravitational attraction is not strong enough to hold the components together. As a result, its stars move away fairly quickly; in a few million years the group is dispersed, and its stars are so scattered in the stellar field that they are no longer recognizable as components of an association. Once again, it follows that the stellar associations we observe today must have formed recently.

All this suggests the idea that star formation[57] is a continuous process that may in fact have been going on since the beginning of the universe.[58] It appears that the stars are formed in groups and that star clusters and associations represent different events of star formation. An educated guess at this point would be that the forges of stars are found in the spiral arms of our galaxy and in the following pages we shall find more than enough evidence to support this hypothesis. As a matter of fact, every astrophysicist today would be ready to swear that it is a well-established, incontrovertible fact.

Table 4 summarizes data on the three types of star aggregates.

·STELLAR POPULATIONS

Now that we have met star clusters, it seems a good idea to construct H-R diagrams for these systems, on the assumption that since the original H-R diagram enabled us to learn many things about the stars in the sun's neighborhood, the same might happen in this case. If we should discover something interesting, we would not only satisfy our curiosity but also impress our fellow astronomers; every new finding, even if due to sheer luck, seems to confer a certain luster on the discoverer—something not to be sneered at. In any case, this is the way research proceeds. A fundamental discovery is always followed by a number of studies that are essentially repetitions of the original work. There is nothing wrong with that. In the first place, fundamental discoveries are not made every day. Furthermore, such follow-up work is needed to substantiate and expand the new finding and to investigate all its possible implications.

To construct an H-R diagram for a globular cluster, we need the spectral classes and the absolute magnitudes of its components. The problem is that it is generally impossible to obtain spectra for these stars; the cluster is such a distant object that we cannot profitably disperse on a spectrum what little light we receive from each individual star. There is a way out of our difficulty, however.[59] Recall that the spectral type is an index of a star's surface termperature and that the termperature determines the star's color. Recalling also what we said about the color index, it is obvious that the spectrum is not needed. Although they do not yield information about spectral lines, observations with color filters give the spectral distribution of the continuum, which is enough for us to determine surface temperature and hence spectral class. Now for the absolute magnitudes. In this case it is not so terrible if the distance of the cluster cannot be measured. The cluster is so far away that all the stars in it can be regarded as being at the same distance from us. Hence the absolute magnitudes of the stars differ from the apparent magnitudes by a constant, which is equivalent to shifting the zero point on the magnitude axis, that is to say, shifting the H-R diagram as a whole from top to

TABLE 4
Principal characteristics of the aggregates of stars in our galaxy

	Globular clusters	Galactic clusters	Associations
Dimensions	50–100 parsecs	<10 parsecs	30–200 parsecs
Mass (in solar masses)	10^4–10^5	10^2–10^3	10^2–10^3
Number of stars	10^4–10^5	50–10^3	10–100 (?)
Color of the brightest stars	Red	Red and blue	Blue
Absolute magnitude (visual)	(-5)–(-9)	0–(-10)	(-6)–(-10)
Star density (solar mass/parsec^3)	0.5–10^3	0.1–10	<0.01
Number of objects observed	125	>1,000	80

bottom. In other words, it makes no difference in this special case whether the H-R diagram is constructed using absolute or apparent magnitudes.

Figure 82 shows the H-R diagram for the globular cluster M 3. In this case the spectral class is plotted against both the absolute magnitude (right-hand scale) and the luminosity in terms of the sun's (left-hand scale). If you compare this diagram with the diagram for stars in the sun's neighborhood, you will notice that they are quite different. The upper part of the main sequence is virtually missing here; after breaking off at spectral class F0, the main sequence branches upward and to the right to join the region of the red giants. In addition, next to the giants, there are several stars that fall in a more or less horizontal branch about absolute magnitude 0. Finally, the region corresponding to absolute magnitude 0 and spectral class A0, which in the original diagram was empty, is now rich in RR Lyrae stars.

These differences might be a peculiarity of the M 3 cluster. It turns out, instead, that this diagram is characteristic of all globular clusters. This fact led to the establishment of two types of stellar populations, one for the spiral arms (Population I) and one for globular clusters (Population II). Analogous studies of the galactic nuclei have shown that they too consist of Population II stars.

Spectroscopic observations have shown that the stars in globular clusters are different in chemical composition from the stars in the spiral arms. Basically, the difference is that Population II stars are deficient in, and may even lack, heavy elements. Some of these stars consist only of hydrogen and helium. This metal deficiency is also observed in the stars scattered in the galactic corona; they too are Population II stars.

One more thing. Population II stars are never associated with gas and dust. In the globular clusters, as far as we know, there is no trace of either gas or dust. Before we elaborate on these observations, let us continue to look around and gather information.

Since we have discussed the galactic clusters, let us construct H-R diagrams for them too. What we find is essentially the main sequence we found earlier in the diagram for Population I stars; hence we are dealing with Population I stars. The diagrams for galactic clusters are not all alike, however; the differences among them involve mainly the extension of the main sequence. In several diagrams the upper part of the main sequence is missing and many stars fall in the region of the red giants. Compare, for example, the diagrams for the Pleiades and for the M 67 cluster, shown in figure 83. As we shall see later on,[60] these differences can be used to determine the "age" of the cluster. For the time being, compare the main characteristics of the two stellar populations, which are summarized in table 5.

It is also interesting to construct H-R diagrams for particular groups of stars. For example, figure 84 shows a combined diagram for the 90 brightest stars and the 82 closest stars.

In passing, I would like to mention that an H-R diagram is a very good way of determining the distance of a cluster. All we have to do is to plot color index, which is also a spectral-type index, against apparent magnitude for the cluster stars. Since the spectral types of main-sequence stars correspond to specific ab-

TABLE 5
Principal characteristics of stellar populations

Characteristic	Population I	Population II
Color of the brightest stars	Blue	Red
Characteristic stars	Blue giants (of large mass)	Red giants (of small mass)
Abundance of heavy elements	High	Low
Age[a]	10^6–10^8 years	About 10^{10} years
Position in our galaxy	Disk	Halo and nucleus
Type of star aggregate	Galactic cluster and association	Globular cluster

a. To be treated in part III.

Figure 82
H-R diagram for the globular
cluster M 3.

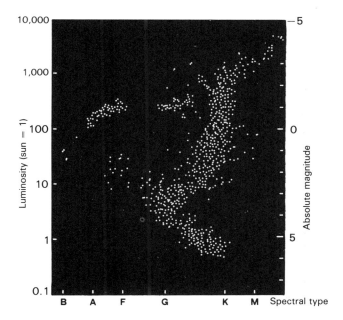

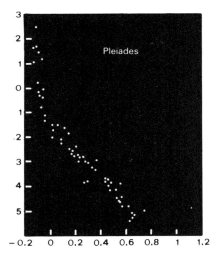

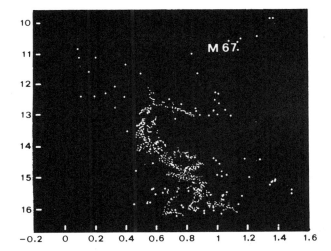

Figure 83
H-R diagrams for two galactic clusters, the Pleiades and
M 67. The former is a
young cluster (5×10^7
years), while the latter is an
old one (5×10^9 years). In
the diagram for the Pleiades
there is only the main sequence; in the diagram for
M 67 the upper part of the
sequence is missing, but
there are many red giants.
The vertical axes for the
diagrams are (left) absolute
visual magnitude and (right)
visual magnitude. The horizontal axis for both dia-
grams is the color index.

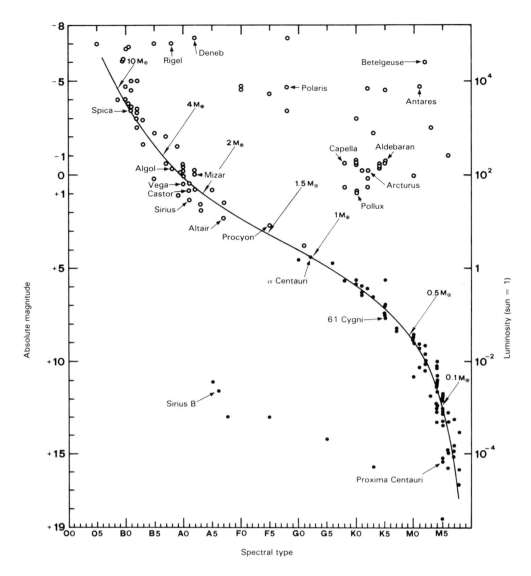

Figure 84
H-R diagram for the 90 brightest stars (o) and 82 stars located within 6 parsecs of the sun (•). Sirius, Altair, and Procyon are included in both groups. In general, the brightest stars are not the nearest ones; they are visible even at great distances because they are intrinsically very luminous. The curve represents the main sequence; the values of stellar masses along the sequence are indicated ($M_\odot$ = 1 solar mass). Observe that the luminosity increases with the mass (mass-luminosity relation); this shows that the correlation between spectral type and absolute magnitude (or luminosity) is also a correlation between spectral type and mass, that is, surface temperature and mass.

solute magnitudes, it follows that for each color (that is, spectral class) the difference between the apparent and absolute magnitudes of main-sequence stars of the same color gives the so-called distance modulus of the cluster. Based on the relation that ties the apparent magnitude (m), the absolute magnitude (M), and the parallax (p), the distance modulus is

$$m - M = -5(1 + \log p) = 5(\log d - 1).$$

Thus once $m - M$ is known, we can find d. An analogous procedure, if you recall, is used to measure the spectroscopic parallaxes of single stars, but in the case of clusters the errors made for single stars are averaged out and the distance can be determined with good accuracy.[61]

• A FINAL LOOK AT THE GALAXY

Before closing this part, I would like to say a few more words about our galaxy. We have already learned quite a few things about it: that our stellar system is a vast spiral; that it is surrounded by a halo containing more than 100 globular clusters and many scattered stars; that there are two distinct types of stellar populations; that star formation is a continuous process; that the stars are formed in groups, or associations, and later dispersed throughout the vast stellar field; and that double or multiple systems are just as common as single stars. We have met supergiants, giants, dwarfs, and objects smaller than dwarfs; stars whose energy fluxes are constant in time, such as the sun, as well as variable and exploding stars. We have learned to determine their distances, masses, sizes, and temperatures. Most important, I think, we have learned that despite its inconceivable vastness, our galaxy can be explored and understood.

We have almost finished the job of gathering observational data. Before we start interpreting them, one more thing needs to be done, namely, to explain why we can confidently state that the Milky Way is one of the stellar systems known as spiral galaxies. As usual, I hope that by explaining how certain conclusions have been reached I shall dispel that aura of ineffable mystery that is so detrimental to our understanding of the world.[62] I shall touch briefly on the methods used in galactic-structure research and some of the most important results.

The interstellar medium makes optical observations of our galaxy very difficult. What we see of it is but a very small part of what we could see without dust. It is the same as looking at a landscape when strong winds raise large clouds of dust; nearby objects can be seen fairly well, but more distant things are blurred or totally obscured. The light that reaches the earth comes from a region no larger than a few thousand parsecs. The brightest portion of the Milky Way is visible at our latitudes, that is, in the Northern Hemisphere, during the summer. It is a little low on the horizon and lies in the direction of the galactic center, which is the densest region of the system.

Red light, as you well know by now, is scattered less by the dust particles. Why not then look at the Milky Way in red light? Better yet, why not look in the infrared? This is precisely what astronomers did as soon as suitable instruments and techniques became available. Infrared photographs, taken with filters and photographic emulsions sensitive to this kind of radiation, have clearly shown the bulge in the galactic center. In addition, observations at increasingly longer wavelengths, in the far infrared, have yielded maps that give the brightness distribution along the Milky Way. By comparing photographs and maps in different wavelengths it is possible to determine the absorption coefficient, that is, the extent to which the dust dims the light and prevents us from seeing the galactic center in all its splendor. It turns out that the visible light from the central bulge is dimmed by a factor of 10^{-10}, or 25 magnitudes.

Yet it is difficult to give up optical observations. Working in the infrared is not that easy. For the near-infrared, at wavelengths of about 1 μm, the photographic plates must be supersensitive—the touch of one's hands is enough to impress them as if they had been exposed to light. For the far-infrared, the equipment is fairly complex and has to work in an environment full of infrared radiation. This is because the instrument itself, the ground, and everything else at room temperature radiates in the infrared. Optical techniques, on the other hand, are relatively simple and well tested. Furthermore, the number of available instruments is far larger. In the visible region of the spectrum, however, the center of our galaxy remains beyond our reach, except when a few gaps in the interstellar clouds allow us to see what lies behind. One of these "windows," with an absorption of only 2 magnitudes, is located 4° below the galactic center. Considering the distance of the galactic center, 4°, or 600 parsecs, is not small—but not large, either, in relation to the sizes involved; we still look in more or less the right direction, and we can certainly see through the central bulge. Among the stars observed in the galactic nucleus are many RR Lyrae variables, which are typical Population II stars.

As its flat shape suggests, our galaxy rotates about itself. If so, we should be able to deduce the motion of the sun in the system from the apparent motion of objects that can be assumed not to share in the general motion of the disk. The globular clusters make good reference objects because their space distribution shows that their collective motion is to some extent independent of that of the disk. All we have to do (obviously it is easier said than done) is to measure their radial velocities. We find that some clusters are approaching us and some are receding, depending on the region of the sky we observe. Our results can be verified by measuring the radial velocities of nearby galaxies, which are certainly not affected by the motion of our own. This type of observation shows that the sun moves at a speed of about 250 km/sec along the galactic plane toward the constellation Cygnus, that is, at right angles to the direction of the galactic center. Since the center is at a distance of about 10^4 parsecs, it turns out that the sun completes a full circuit in 250×10^6 years. With these data the mass of our galaxy can be quickly estimated. Let us assume that all its mass is concentrated in the nucleus and disregard the mass that is farther away from the center than the sun. What this means is that the actual mass of our galaxy will be somewhat larger than our calculations show. According to Kepler's third law,

$$\frac{a^3}{P^2} = \frac{G}{4\pi^2} (m_1 + m_2),$$

where a is the semimajor axis of the orbit, P the period of revolution, G the gravitational constant, and m_1 and m_2 the two masses in the system considered. Writing the law first for the galaxy-sun (G-$\odot$) system and then for the sun-earth ($\odot$-E) system gives

$$\frac{a_\odot^3}{P_\odot^2} = \frac{G}{4\pi^2} (m_G + m_\odot)$$

and

$$\frac{a_E^3}{P_E^2} = \frac{G}{4\pi^2} (m_\odot + m_E).$$

If we take the semimajor axis of the earth's orbit as the unit of measure (that is, $a_E = 1$ astronomical unit) and 1 year as the unit of time (that is, $P_E = 1$), and if we treat m_E and $m_\odot$ as negligible in comparison with, respectively, $m_\odot$ and m_G, then dividing the second equation into the first yields

$$\frac{a_\odot^3}{P_\odot^2} = \frac{m_G}{m_\odot} .$$

Take $m_\odot$ (1 solar mass) as the unit of mass and recalling that 10^4 parsecs is about 2×10^9 astronomical units; then

$$m_G = \frac{a_\odot^3}{P_\odot^2} = \frac{(2 \times 10^9)^3}{(2.5 \times 10^8)^2} ,$$

which is about $8 \times 10^{27}/6 \times 10^{16}$, or 1.5×10^{11}. The mass of our galaxy is thus equivalent to at least 1.5×10^{11} solar masses, or roughly 100 to 200 billion suns. Since 1 solar mass is roughly the average mass of the stars, we can conclude that our galaxy contains at least 100 to 200 billion stars, probably more.

We might also wish to find out how the sun moves in relation to the stars in its neighborhood. Recall that the space velocity of a star is defined as the velocity of the star in relation to the sun regarded as stationary. We could as well consider the star stationary and the sun in motion, since what we are really talking about is the *relative* velocity of the stars with respect to the sun. But we can take as reference system a number of stars whose average motion is nil and determine the sun's motion in relation to the system. This is done by measuring the proper motions and radial velocities of many stars in the sun's neighborhood (that is, within about 100 parsecs of the sun). Without going into the actual reasoning and calculations, it turns out that in relation to this local reference system the sun moves approximately at a speed of 20 km/sec in the direction of the star Vega in Lyra. The point toward which the sun moves is called the solar apex. Measurements of radial velocities show that the stars near the apex move on the average toward the sun, while the stars in other regions move on the average away from the sun, the more so the farther away they are from the apex. The stars near the antiapex appear, of course, to recede at the greatest velocities, which are equal and opposite to the velocities of the stars near the apex.

Note that the velocity of the sun in relation to the local system is small. What is measured, in effect, is the relative velocity of the sun with respect to the system, that is, in galactic terms, a residue of the galaxy's differential velocity. What this means is that the galaxy does not rotate as a rigid body; the inner part moves with a higher angular velocity than the outer part.[63] (The same thing happens in the solar system; Mercury's angular velocity is higher than Venus's, which in turn is higher than the earth's, and so forth.) This effect is demonstrated by measuring the velocities of stars in different regions of the galaxy. Stars closer to the nucleus precede the sun, while stars farther away from the nucleus than the sun lag behind. Naturally this effect is the more conspicuous, the farther away the stars are from the sun in either direction. Figure 85 illustrates the velocities of rotation of the various parts of our galaxy as a function of their distances from the galactic center.

Incidentally, most of the stars in the sun's neighborhood have velocities that cluster around 20 km/sec. They are known as low-velocity stars. The mass of interstellar gas and dust moves, as a whole, in the same manner (as shown by measurements of the Doppler shift of the lines of emission nebulae and of the lines due to interstellar absorption). Thus it can be concluded that aside from small effects due to the differential rotation of our galaxy, the block that constitutes the sun's neighborhood revolves about the galactic center as a whole, with a velocity of about 250 km/sec.

In the sun's neighborhood, however, there are stars that move at speeds of 70 km/sec or more—the so-called high-velocity stars. In this case, too, we must keep in mind our reference system. These velocities are high from the point of view of the sun. But this has nothing to do with the velocity in relation to the center of the galaxy. When I am driving a car my passenger has zero velocity with respect to me. But for the gas station attendant who watches the car go by, my passenger has a high velocity. And the person who rides his bicycle along the highway has a higher

Figure 85
Differential rotation of our galaxy. The velocities of rotation of the stars about the galactic center of mass vary with their distances from the center. The gas and dust scattered in the galaxy share in this motion.

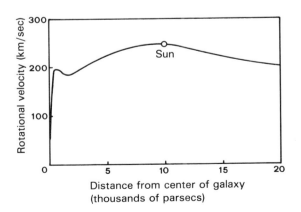

velocity, for me, than my passenger, but a much more modest one for the attendant. The same is true for high-velocity stars, which in the general motion of the galaxy are generally slower than low-velocity stars. They are believed to be stars that travel highly eccentric galactic orbits and therefore pass through the sun's neighborhood (see figure 86). Furthermore, their orbits are generally inclined with respect to the galactic plane; hence they cross it. For these reasons we are often dealing with stars of the galactic halo.

In all probability, as they revolve about the galactic center the globular clusters also cross the equatorial plane (twice per revolution). What a catastrophe, you will say. As it passes through the disk, a globular cluster containing millions of stars must create absolute havoc; imagine all those stars bumping into each other! Instead, there are no hits—or, rather, they are highly improbable—for the simple reason that interstellar distances are enormous. Think of two groups of people crossing each other at an intersection, with the difference that the individuals are far, far closer than the stars.

To close, I mention briefly the radio observations at the 21-cm wavelength. As I said earlier, this emission is due to the (forbidden) transition taking place between the two levels of the hyperfine structure of the fundamental level in the neutral hydrogen diffused throughout our galaxy. Radio waves do not interact much with the interstellar medium (are not absorbed to any significant degree) because their wavelengths are very large compared with the obstacles that might cause them to be absorbed (dust grains and atoms). A useful analogy would be the cases when a sea wave meets a sea wall and when it meets a pole buried in the sea bottom. In the first case, the wave is blocked, and on the other side of the wall the sea is calm; in the second case, the wave passes around the obstacle, which is not large enough to block it. Similarly, radio waves travel fairly freely in space and pass easily through the interstellar medium, which, in contrast, is quite opaque to visible light.

The 21-cm radiation is produced by clouds of neutral hydrogen. If the source (a cloud) moves along the line of sight, the 21-cm line will show a Doppler shift. Thus from observations of this line we can deduce the motion of the gas cloud. Conversely, if we know the velocities of the various parts of our galaxy as a function of their distances from the center, we can find the distance of a particular cloud from the center.

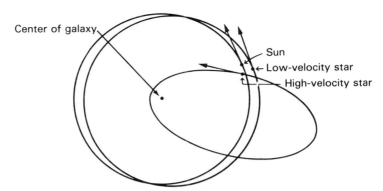

Center of galaxy

Sun
← Low-velocity star
← High-velocity star

Figure 86
The galactic orbits of the sun and a so-called low-velocity star, which actually means a star of low velocity with respect to the sun. (An analogy would be the case of two trains traveling in the same direction on parallel tracks at great, but slightly different, velocities—say, 100 and 103 km/hour. A passenger in either train would measure a low relative velocity for the other train.) The orbit of a high-velocity star is also indicated. In this case, too, one should say high velocity with respect to the sun. The actual velocity of this star in our galaxy may be lower than that of the sun.

Observations of the 21-cm radiation also show the distribution of neutral hydrogen in our galaxy. At first sight it might seem impossible to separate the signals that come from different clouds in the same line of sight. Two stars in the same line of sight appear as one because the nearest one hides the other. Thus a cloud should hide the cloud, or clouds, that happen to lie behind. But if galactic clouds share in the differential motion of the galaxy, it follows that the farther a cloud is from the galactic center of mass the slower it moves, and that the closer it is the faster it moves. As a result, the radial components of the velocities in relation to the sun are different for different clouds; hence, the Doppler shifts of their 21-cm line will be different. Radiation from all points in the line of sight reaches the radiotelescope at the same time, but is automatically separated by virtue of the Doppler effect.

Figure 87 illustrates a scan of the sky at the 21-cm wavelength. The peaks correspond to gas clouds moving at different speeds. Each cloud, of course, emits at the wavelength of 21 cm, but owing to the Doppler effect the observer receives signals at different wavelengths for different clouds. Figure 88 is a radio map of neutral galactic hydrogen obtained from observations of the 21-cm line. The spiral structure is quite evident; so, you see, we can tell the shape of our galaxy even though we are inside it. There is no question that it is a spiral galaxy with several arms. In one of the spiral arms, at the periphery, there is the sun, the earth, you and I.

We have certainly come a long way from Galileo's time. We used to be the kings of creation, the center of the world, the hub of the universe. Now we are nothing. That's life. I leave it up to you to draw your own conclusions. According to your temperament, you may take it philosophically, turn it into a joke, or

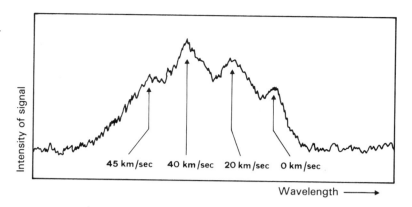

Figure 87
Emission from neutral galactic hydrogen at the wavelength of 21 cm. If the velocity of the gas were the same as that of the observer, there would be only one peak, corresponding to the velocity 0 km/sec. But the various spiral arms move at different speeds (with different radial components), and although the emission is at $\lambda = 21$ cm everywhere, because of the Doppler effect the observer receives signals at slightly different wavelengths.

consider it a great tragedy. On the other hand, you can dream up any world you like and give humanity any role that seems best to you.

I realize that some of the celestial phenomena may be quite awe-inspiring, but it is only because we are not used to living with nature, or at least a certain kind of nature. For me, the rotation of our galaxy or a bright nebula are no more impressive than a falling rock, the existence of flowers, or the play of color at sunset. Nothing is overwhelming and everything is. No doubt, it is a question of habit. We are so used to flowers and sunsets that we do not pay them much attention. If our planet had two suns, each a different color, we would get up in the morning and catch the bus without giving them a second look.

Yet I know that for many people celestial phenomena are still a source of great awe. The reason is that astronomy deals with ideas that for a long time have frightened people—the immensity of creation, the smallness of man, the terrible justice of the creator, the infinity of time, the short span of human life, our nothingness in a universe we cannot grasp. I cannot help feeling that this is the wrong way to look at the sky. There is enough fear—or happiness, or wonderment—here on earth that we do not need to importune the heavens.

Why should I mind that the sun is at the outskirts of town? It may not be the most prestigious location, but personally I like it fine. I would not care to live on a planet with two suns or in a globular cluster where I could see the whole galactic disk on a skyorama. I am content to be in a spiral arm with my own sun and earth. I have everything I need here—colors, sounds, starry nights, storms, waves, friends, work, trees, happiness, sadness, light, air, and water.

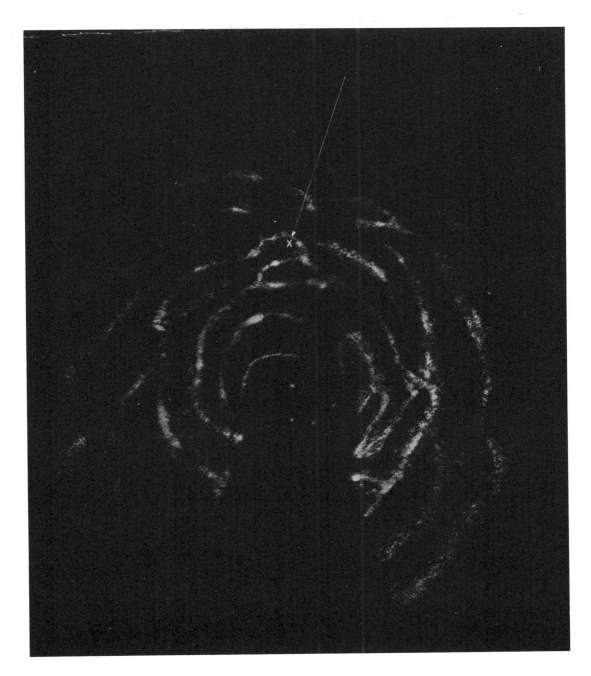

Figure 88
Radio map of our galaxy obtained from observations in
the 21-cm line performed at
Sidney, Australia, and Ley-
den, The Netherlands. The
arrow indicates the position
of the sun in the galaxy.

You must be wondering whether I have written this whole book simply to tell you all my likes and dislikes. I have not, but I am tempted to go on musing about things. I am as much of a daydreamer as the next man. Enough rhetoric, however, and back to reality.

We have seen how astronomers use their brains to ask scientific questions, and it is pretty amazing to think that a small creature on this troubled planet has been able to probe our galaxy and to know it better, in certain respects, than the earth itself. All the facts we have gathered must be interpreted now. The fact is, there is no science without interpretation. Only by explaining the facts, by producing a model, a theory, a formula, can we ever say that we have understood how things work.

Understood? Now quite, perhaps. Even though it proceeds from known, certain facts, every interpretation is somewhat arbitrary in the final analysis. I wish to stress once again that science has a history, like any other human activity. We are interested in all the things that in one way or another relate to us. Astronomy is the study of the relation between us and creation. We want to know how the universe is made and works, in part, also, to understand what we can expect from it, whether it can be of use to us (in a broad sense), how it can affect us. The universe is a giant that has frightened man and inspired religions for thousands of years. It must be understood. Once explained, the facts will no longer be frightening. On the contrary, they will be as obvious as rain, snow, or the dawn of a new day. As even the dumbest dictator knows, knowledge is freedom.

PART III
STELLAR EVOLUTION

Is this raving? I ask myself

sometimes, aghast not to know,

aghast at my bounds,

aghast at the earth, this speck

that coalesced amidst cosmic gusts

in a time out of mind—

only to dissolve perhaps

into nameless maëlstroms.

Living horrifies

              and fascinates me.

We are tiny microbes

absently poised

between arrogance and despair.

Danilo Dolci
From "The Lord of the Ticks"

• A BRIEF INTRODUCTION

This last part will probably be more difficult for you to understand and for me to write. I do not mean to imply that everything else was all that simple, even though a certain professional perversity should lead me to say so. You know how it is. Few scientists care to admit that they have troubles understanding something or, conversely, that there are things hard to understand. It is a very convenient attitude, particularly when they have to explain something, in that it frees them of many responsibilities. I have tried to rid myself of this professional bias, and in the previous chapters I was helped by the fact that the subject matter—half-facts, half-interpretations—could be treated fairly yet imaginatively. Now I find myself in trouble. What I have to say—and it must be said not to leave the picture unfinished—requires a knowledge of physics that goes beyond what is taught or could be taught in high school. Very likely, I will not be able to be as plain as I hope I have been, and this will make for problems in comunication and an occasional shift in style. In all honesty, let me say that if you happen not to understand something, it is not because you are mentally deficient, but because of the difficulty of the subject and my shortcomings as a teacher. If the following pages seem unconvincing or unsatisfactory, I hope you will feel motivated to do some more reading.

On my part, I promise not to hide my inability to write clearly behind stock phrases like "and by easy steps it follows. . . ." On your part, I expect some cooperation and a little effort to try and understand what I may (inadvertently) not explain too well.

• EVERYTHING HAS A BEGINNING AND AN END

If you look around, you will not find anything that lasts forever. Even things that embody the idea of permanence and immortality, such as the mountains or the oceans, are neither unchanging nor eternal. Today we know this for a fact. But even the ancients sensed that it had to be so, that nothing could have existed forever and that everything must have had a beginning. The question was how. As a working hypothesis—to explain natural phenomena as frightening as they were mysterious—they invented one god, two gods, hundreds of gods, angels and devils, good and evil spirits, and they invested the stars, planets, sun, moon, and comets with supernatural powers. Religion, mythology, astrology, and magic were born—the offspring of ignorance, fear, and unhappiness. In the light of reason they can only be regarded as humanity's first attempt to explain the world.

As we all know, however, it is not entirely past history. People are still afraid, and although intrigued by a certain type of knowledge, they choose to ignore it instead of using it to attain intellectual freedom. They are interested in science, and astronomy in particular, but only for the sake of acquiring information that, in itself, is not very useful and, if anything, tends to create fear of the ineffable mysteries they think they have glimpsed. The latest astronomical findings are loudly acclaimed as extremely interesting, fascinating, wonderful, unbelievable, and that is the end of it. Instead of drawing their own conclusions, people continue to practice rituals that go back directly to our ignorant and frightened ancestors. There seems to be no other explanation for the survival of magic practices that turn adults into children afraid of the dark or the big bad wolf.

And the children are many, as demonstrated here in Italy by the fact that our nonreligious politicians feel they have to bow to the leaders of what they can only regard as a superstitious sect because of the power these people wield over the large mass of believers. Superstition is taught in Italy's schools. Italy's state-run radio and television regularly feature Catholic priests, rabbis, protestant ministers, and astrologers—preferably, though for different reasons, the first and the last.

Why should we live in fear and superstition? Why are we taught to be afraid? Never mind. I should not let myself be carried away by feelings of indignation that are ineffectual and, all told, a little naive. The battle against superstition and prejudice has been going on for a very long time, with no end in sight, and it will take more than a few words to win it.

Even in the past, however, not everybody believed that the world started with an act of creation. Some thought that it might be eternal. But in any case, I think everybody agreed that each individual thing had a finite life, that everything must have a beginning and an end.

Let us reason in modern terms and consider the sun. We know that it has been there for a very long time. And we know that it keeps on radiating a huge amount of energy. There is no doubt, therefore, that the sun is consuming itself, day after day, like a burning log. Since it has no other source of energy than itself, it follows that at a certain moment the sun will come to an end.

Working backward, it follows that unless it was infinitely large, the sun cannot ever have been what it is today. It must have started somehow. It does not take much imagination or thought to conclude that the same must be true for every other star, our galaxy, and perhaps the universe as a whole. The reason why we cannot see changes on a cosmic scale is that our lives are too short. Humanity is a latecomer in the history of our planet. The earth formed approximately 5 billion years ago. The first traces of life date back 3.5 billion years, but human beings appeared only about 3 million years ago. In mathematical terms, 3×10^6 is to 5×10^9 as 6×10^{-4} is to 1. In other words, humanity's time on earth is only 0.0006 of the earth's life, which is like 50 sec in a day. Recorded history is 10,000 years old, which corresponds to 0.17 sec in a day, and we have been studying the sky for 400 years, which corresponds to 0.007 sec in a day. It is evident that in such a short time very few changes can be observed and that the sky always looks the same. During a person's lifetime nothing at all happens; the sky is imprinted on the celestial vault, and any change in the pattern is an illusion created by the earth's motion. It took generations to observe some of the phenomena I have described, and some things would have remained unknown forever if it had not been for instruments that have immeasurably expanded human vision. Even so, it is impossible to observe significant changes in the individual stars and to hope to formulate a theory of stellar evolution based on direct observation. We can see a plant, or a fly, being born, growing, and dying, and a few generations of researchers can gather enough data to deduce the fundamental elements in their life cycles. But the stars? We suspect—indeed, we know—that they must be subject to evolutionary changes for the simple reason that they will surely come to an end and hence must have had a beginning. No star, no matter how it started or how it will end, can remain exactly the same during its entire life. But how can we see these changes?

In principle, at least, the problem of investigating stellar evolution is not as hard as it seems. To explain how it can be done, I shall use an analogy I read somewhere and liked very much.

Suppose we know nothing about trees and we have a few hours to spend in woods untouched by human hands and governed solely by natural laws. The woods are the sky, and the few hours represent the short time humanity has had to study the unknown world of stars. In a few hours we cannot hope to see a tree being born, growing, becoming old, and dying. But if instead of wasting our time looking at a single tree we observe how the woods came into being, then we may be able to understand the trees' life cycle . of birth, growth, aging, and death.

In the first place, we notice that the trees can be divided into groups according to appearance. We also observe that each group contains specimens of different sizes and that the largest trees in each group are about the same size. Finally, we observe that some of the largest trees have dried up and perhaps fallen down, whereas the smallest trees are green and tender. At the end of our exploration we can produce a reasonable theory to explain what we have seen. Trees are born as small things; by mechanisms that are as yet unclear they grow to a maximum size, and after a time they die.

The same type of reasoning can be applied to a group of human beings. In this case we could easily deduce that a human being is born small, grows fairly rapidly to the maximum typical size, remains in this condition for a period of time, and then dies. This is the way to study the galactic population.

· THE H-R DIAGRAM REVISITED

It comes fairly naturally at this point to recall the classical H-R diagram (see figure 43). Since it concerns all of the stars, or at least a large number of them, it must hold a clue to stellar evolution. Why is it that most stars lie on the main sequence? A reasonable assumption is that so many stars are found there because they spend a good deal of their lives in the physical conditions characteristic of the main sequence. It is not by chance that if you take 100 people at random, most of them turn out to be adults; since people spend most of their years as adults, there are many more adults around than children.

Thus we shall proceed on the assumption that the main sequence is the adult stage of a star's life. If this assumption is correct, the pieces of our puzzle should fall neatly into place. But if it does not work out, we shall have to throw it out, intriguing though it may be, and start all over again.

Another significant feature of the classical H-R diagram is that the fainter stars, except a few (white dwarfs; lower left in figure 43), are well aligned along the main sequence, whereas the brighter stars are scattered all over the top. Why is that? And what do we make of the H-R diagrams for globular clusters and open clusters?

In the former the main sequence stops about halfway and the rest of the stars branch off to the right; in the latter the star population is distributed almost exclusively along the main sequence. Why? To find the answers to these questions, we must first understand the nature of a star.

•THE SOURCE OF STELLAR ENERGY

In essence, a star is an object that radiates energy. To understand the life of a star we must therefore understand the mechanisms of energy production. A star is made of matter. So is a stone and so are we. But a stone does not produce energy, and neither do we. How does a star do it?

I am sorry I cannot give you an exhaustive explanation. It would require a good deal of modern physics to do it, and this is truly beyond the scope of this book. I can only give you the gist of it, and if you are not satisfied, there are plenty of books that deal with twentieth-century physics.

Atomic nuclei consist of particles, known as nuclear particles. The most common nuclear particles are the protons, which carry the smallest electric charge, and the neutrons, which are devoid of charge. Protons and neutrons have an almost identical mass (1.66×10^{-24} g), the neutron being slightly heavier. The mass of the nucleus is essentially given by the sum of the masses of its protons and neutrons. Practically, this is the mass of the atom because the electrons, which are also part of the atom (not the nucleus), have a mass that is about $1/2,000$ that of the protons. To put it another way, it takes about 2,000 electrons to make the mass of a proton.

The simplest element is hydrogen, whose nucleus consists of 1 proton. The nuclei of the other elements are made of different combinations of protons and neutrons. The helium nucleus, for example, consists of 2 protons and 2 neutrons, the lithium nucleus of 3 protons and 4 neutrons, and so forth. The mass of a composite nucleus, however, is not exactly the sum of the masses of its particles, but slightly smaller. This difference is called mass defect and is greatest for the iron atom. The nucleus of an element can be made either by combining (fusing) the nuclei of lighter, simpler elements or by splitting up the nuclei of heavier, more complex elements. These processes are what we call nuclear reactions. But the total mass of the product of the reaction is not the same as the original mass. This difference in mass is equivalent to the difference between the sum of the mass defects of the nuclei involved in the reaction and the sum of the mass defects of the nuclei resulting from the reaction. Einstein demonstrated that mass is a state of energy (that is, mass and energy are equivalent) and that the mass that is missing at the end of a nuclear reaction has been converted into energy according to the equation $E = mc^2$, where E is the energy resulting from the "disappearance" of the mass m and c is the speed of light. This is the basis for the theory, formulated in 1928, that the source of stellar energy resides in nuclear reactions in which light nuclei fuse to produce heavier nuclei with release of energy.

We have learned that the stars consist essentially of hydrogen and, to a lesser extent, helium. Obviously, these two elements— of all elements the lightest and simplest—must be the main ingredients of stellar nuclear reactions.

It takes 4 hydrogen nuclei to make 1 helium nucleus. The latter has a mass of 4.00389 atomic units, while the mass of the hydrogen nucleus is 1.00813 atomic units. A quick calculation shows that 0.02863 atomic unit is missing from the product of the reaction; in other words, 0.71% of the mass of the 4 hydrogen nuclei has been lost. This means that for 1 g of hydrogen that turns into helium, 0.0071 g is converted into energy, and the amount of energy released is

$$E = 0.0071 \times (3 \times 10^{10})^2 = 6.4 \times 10^{18} \text{ ergs.}$$

Knowing the sun's luminosity (the energy radiated by the sun every second), which is 4×10^{33} ergs/sec (or 4×10^{26} watts), we can quickly calculate how much solar hydrogen turns into helium every second. Dividing 4×10^{33} by 6.4×10^{18}, we get 0.63×10^{15} g, or about 600×10^6 tons. If this is the mechanism of energy production, 600 million tons of the sun's hydrogen turn into helium every second, and 0.71% of this mass, or 4 million tons, leaves the sun every second in the form of energy. (The reason why astronomy deals with such large numbers is that celestial phenomena are so enormous; just how enormous, you realize as soon as you make some calculations.) The idea that the sun must lose 4 million tons of its mass every second to radiate as it does is almost too much to accept. Let us go on, however. How long would it take for one-tenth of the sun to turn to helium at the rate of 600 million tons per second? Or, what is the same thing, how long does it take for 0.71% of one-tenth of the solar mass to be converted into energy at the rate of 4 million tons/ sec? The mass of the sun is about 2×10^{33} g. One-tenth of this mass is 2×10^{32} g. To consume all of them at the rate of 6×10^{14} g/sec takes $2 \times 10^{32}/6 \times 10^{14} = 0.3 \times 10^{18}$ sec. Since there there are 3×10^7 sec in a year, it comes to about 10×10^9 (10 billion) years. In all this time the sun will only lose 0.07% of its mass in the conversion of matter to energy. Which tells you how truly enormous the sun is.

Hence it is quite possible for nuclear reactions to be the source of stellar energy. No need to look for another mechanism; the one we found is more than sufficient. Although we appear to be on the right track, we must pursue our idea a little further. Specifically, since there is nobody there to kindle nuclear reactions, they must start by themselves. No sooner is a problem solved than a new one appears! Now we have to figure out whether inside a star are to be found the right conditions for starting nuclear reactions in which hydrogen nuclei combine to form helium nuclei. And if so, we then must figure out which of these nuclear reactions are possible, and after that, which of these possible reactions suit our purposes. The problem is far from simple. To begin with, for a composite nucleus to form within a star, the component nuclei must meet. This precondition cannot be taken for granted because it depends first of all on the density of stellar gas. If the gas is very tenuous, one could wait till Kingdom come for such encounters. Hence there is a lower limit to the density below which nothing happens. In addition, the nuclei must collide with enough speed to overcome the repulsive forces that tend to keep them apart; and the speed of the particles depends on the temperature of the gas.

In the second place, one must take into account the size of the nuclei; the larger the nuclei, the easier the collisions. Actually, speaking of "size of the nucleus" is a bit crude, but it will do for a simpleminded model. The correct expression is "effective cross section"; the same particle has different sizes, so to speak, for different kinds of reactions.[64] Finally, once a new nucleus has formed from the collision of two nuclei, it may remain as is or it may break up into other nuclei different from the original two. In sum, we need to know the probability that nuclei will combine and break up under certain physical conditions.

Fortunately, we have nuclear physicists to take care of this kind of thing. On the basis of experimental data and theories, they come up with the formulas that the astrophysicists need to determine what may happen in a star, or in different parts of a star, given a certain chemical composition, temperature, and density. To bring to a close this purely physical digression, I should mention that there are two kinds of reactions whereby helium nuclei can form from hydrogen nuclei. One is the so-called carbon cycle. In this reaction, which is fairly complicated, 4 hydrogen nuclei go to make up 1 helium nucleus. The carbon acts as a catalyst and at the end is regenerated. This process is efficient at very high temperatures (more than 15 million degrees kelvin). The other mechanism is the so-called proton-proton reaction: A proton combines with another proton and then the product combines with another proton. Subsequently, the reaction can go in either of two different ways, with the same result. In one case, 2 particles, each consisting of 3 protons, combine to make helium (plus 2 protons). In the second case, a nucleus containing 3 protons combines with a nucleus containing 4 protons to make a beryllium nucleus, which is then converted to lithium; finally, by combining with another proton, the latter makes 2 helium nuclei. The proton-proton reaction is more efficient at lower temperatures (in a manner of speaking—we are still talking of temperatures on the order of a dozen million degrees).

·THE INTERIOR OF A STAR

As we have just learned, the central temperature of a star must be at least 12 million degrees to produce the nuclear reactions that convert matter into energy. The energy liberated in these processes consists of photons of very short wavelengths, namely, γ rays and x rays, which do not leave the star, but are absorbed by the overlying material. It is due mainly to this effect that the overlying material is heated to a temperature that determines its emission. Energy then flows outward from this material, thereby additionally heating up an ever larger volume of gas.

The temperature declines steadily from the center of the star to the surface. In the case of the sun, you may recall, the surface temperature is slightly less than 6,000°K. Now we must work our way backward through the various layers of the star to see whether it is reasonable to assume that the temperature and density at the center are the right ones for the occurrence of thermonuclear reactions. Unfortunately, we cannot do it here because we would have to wade into a sea of formulas that would be out of place in this book. What we can do is to follow the reasoning along its general lines.

First of all, we must accept that the laws of nature are valid not only on the earth but on the stars as well. This is not too much to ask. Our knowledge of the physical world has been put to the test so many times without disappointing us that we have no reason to doubt its validity. Thus we can safely assume that Planck's law, the law of gravitation, and the laws governing the absorption and emission of radiation and nuclear reactions obtain within the stars. On this theoretical basis, and starting from surface conditions, which are well established experimentally, we extrapolate to the interior and build a model of a star that is consistent with our observations.

Generally speaking, we must determine how mass is distributed through a star (we cannot assume that it is uniformly distributed); what is the chemical composition of a star's various layers; and how density, temperature, and pressure within a star vary with depth. Every layer will turn out to have a particular set of physical conditions, which, according to the laws of physics, will determine how it emits and absorbs radiation. The stellar gas will be differently ionized at different depths, because of associated temperature and pressure differences, and will be subject to a gravitational force (which varies with the distance from the center) that tends to make it collapse inward. At the same time, the pressure of the gas itself, along with radiation pressure, tends to make it expand. In those regions of the star where these forces balance each other, the gas will be stationary; but if there are imbalances, the gas will tend to move, and convective currents will form. Our model must take all this into account and more: such things as acoustic waves, electric and magnetic fields, stellar rotation, and so forth. It must take into account all the possible interactions between the various parts of a star and between matter and radiation. And once this is done, we must make sure, again on the basis of known laws, that the star we have built behaves like an actual star in front of our eyes. Our model star must emit the same spectrum as the one we observe by means of a spectroscope. It must be a continuum of the same shape and intensity; it must have the same lines (same elements, same number), at the same wavelengths; and these lines must have the same character (strong, weak, broad, narrow).

All told, it is no small thing we ask from the theoreticians. Indeed, it is such a complicated problem that only today, thanks to modern computers, can it be readily tackled. Of course, it does not have to be solved all at once. One can start with a rough model and then proceed by trial and error. Or one can concentrate on particular phenomena (electric fields, for example) if they appear to play a crucial role. Even so, it is a mammoth job.

Some fundamental conditions must be taken into account in determining the structure of a stable star, such as the law of perfect gases, thermal equilibrium, and hydrostatic equilibrium.[65] Also, one needs to determine the mechanisms of energy transport (from one part of the star to another, from the center to the surface), the capacity of matter to absorb radiation (which is called opacity), and the rate of energy production in the stellar core. These quantities are related by equations whose solutions yield a table that gives such things as temperature, pressure, and density at various depths, as well as other information needed to calculate the spectrum of the model star. (In fact, such a table *is* the model.)

In general, a description of the physical conditions prevailing in the interior of a star involves finding equations that express four quantities—mass M, temperature T, pressure P, and luminosity L (the energy emitted every second)—as functions of the distance from the center of the star. The mass and luminosity when the distance is the radius are, of course, the total mass and luminosity. The temperatures and pressures of the outermost layers are obtained from observations and theories of stellar atmospheres; but to a first approximation, when only the macrostructure of the star is considered, they can be assumed equal to zero given the very high temperature and pressure in the interior.

From the conditions on its surface, the basic equations give the values of M, T, P, and L for a star's underlying layer. From these, using the same equations, we find M, T, P, and L for a yet deeper layer. And so on down to the stellar core. This method has given us models for the interior of stars—the sun in particular. The first models of the sun are more than 40 years old. The most up-to-date ones are very consistent. Their results for the sun's core are as follows: $T = 15 \times 10^{6}$°K, $P = 3.4 \times 10^{11}$ atmospheres, ρ (density) $= 160$ g/cm^3, and $X = 0.38$ (hydrogen content).

Figure 89 shows the distributions of the mass and energy production in the sun's interior. Observe that 80% of the sun's mass is concentrated in a sphere whose radius is less than half the solar

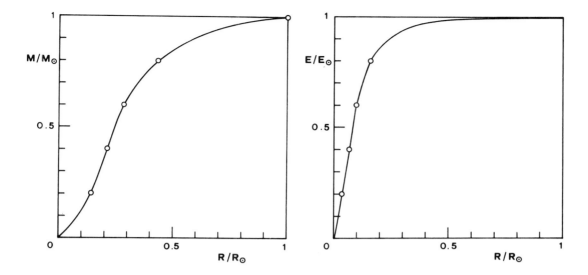

Figure 89
Distributions of mass (left)
and energy production
(right) in the interior of the
sun. $R_\odot$ is the sun's radius;
R, the distance from the
sun's center; $M_\odot$, the sun's
mass; M, the mass of the
sun from its center up to
distance R; $E_\odot$, the sun's
total energy production; E,
the energy production of
the sun from its center up
to distance R. (These definitions lead directly to the following inequalities: $0 \leq$
$R/R_\odot$, $M/M_\odot$, $E/E_\odot \leq 1$.)

radius and that the production of 80% of the total energy emitted
by the sun occurs within a sphere whose radius is one-sixth the
solar radius. It is in this region, very dense and hot, that the energy
is in fact produced.

We can conclude that in the interior of the sun exist the right
conditions for the start of nuclear reactions. We can also be sure
that they are actually kindled because our model works only if
we assume that 95% of the sun (by mass) consists of a mixture
of hydrogen and helium (two-thirds hydrogen and one-third helium)[66] and 5% of the other elements, that is, if we assume the
sun to be a sphere of hydrogen with some helium and traces of
the other elements—which is precisely what observations tell us.

Once we have learned the trick, we can make all kinds of
different models by combining chemical composition, temperature,
density, and anything else we want, including age. Let us elaborate
a little on the question of age. I can make a theoretical model of
the star as it is now. But I can do more. I can think that the star
did not always exist, but that at a certain time (age zero) it formed
somehow, from a certain mixture of chemicals. I can then start
the process of evolution from there. The complications are enormous, of course, but suppose I have done all of this. Let us assume,
then, that at age zero the star begins to work and that nuclear
reactions come into play. How will the star evolve? The chemical
composition, for one, must surely change because some elements
combine to form others. After working for a given number of
years, my model star must end up as the star I observe today.

Not at all simple, believe me. But sooner or later somebody is bound to make a fundamental breakthrough, and everybody else follows in his footsteps. A small change here, a small change there, let us add this, take away that, and see what we have. It becomes almost a game, and if not really a game, at least a routine. If you then add the modern computer, with its tremendous calculating power, the game can go on until a model is found that fits the observations.

Today we know fairly well how a star works. Although the models are theoretical, everything works out fine, on the whole, and the results show that our theories are essentially correct. In other words, it is pretty certain that our views correspond to reality. Of course, it cannot be excluded that an unexpected finding might force astronomers to revise and to modify the picture, but it is extremely unlikely that it will have to be done over.

According to the Russell-Vogt theorem, if a star is in hydrostatic and thermal equilibrium and if nuclear reactions are the only source of the energy it radiates, then the structure of the star is entirely determined by the chemical composition of the star and its total mass. This means that stars with the same chemical composition may have different characteristics, yes, but these characteristics depend solely on one parameter—mass. The luminosities of these stars, for example, vary according to their masses. Thus if we make a graph with the stars' masses plotted against their luminosities, we shall find a mass-luminosity relation, which is the relation we discussed earlier. And if we plot spectral types, that is, temperatures, against luminosities, we find the main sequence of the H-R diagram; hence the main sequence is the locus of stars of the same chemical composition but different masses. By the mass-luminosity relation, the most massive stars are the hottest and most luminous; they are blue and lie on the upper part of the main sequence. Stars of smaller mass have lower surface temperatures and are less luminous; they are red and fall on the lower part of the sequence. The H-R diagram in figure 84 shows the masses for various parts of the sequence.

The fact that most of the stars as well as the sun lie on the main sequence means that the chemical composition of most of the stars is about the same as the sun's. (Herein lies the importance of studying the chemical composition of the sun.) The fact that the main sequence is a belt rather than a line indicates that there are small differences in chemical composition among main-sequence stars.

And what about the stars that do not fall on the main sequence? Either they differ in chemical composition, or they draw their energy in part or exclusively from processes other than nuclear reactions, or they are unstable stars to which the Russel-Vogt theorem does not apply.

And what about the length of the main sequence? Is there a limit to it, at either end, or does the sequence go on indefinitely? In other words, are there upper and lower limits for stellar masses, or can a star have any mass at all, however large or small? The answer is that there seems to be a limit at each end, although not well specified. The problem is a little simpler in the matter of the lower limit; it appears that an object with a mass less than one-hundredth the solar mass cannot become a star because such a mass is not sufficient to produce (in its core) the pressures and temperatures needed to start nuclear reactions and the production of energy. Such an object is destined to be a dark body, perhaps the planet of a true star. It is more difficult to set an upper limit to stellar masses;[67] but in fact no star or stellar object of more than 60 solar masses has been observed or identified.

• THE FORMATION OF A STAR

At this point we can begin to formulate a theory of stellar evolution. A star cannot have existed forever for the simple reason that were it eternal, it would have to be an infinite reservoir of energy. To the best of our knowledge, a star radiates energy by burning its own material in a process that converts mass into energy. Since the mass of a star is not infinite, either our physical laws are rubbish or the star must come to an end, sooner or later, one way or another.

Once we agree that a star must have a beginning, the question is, How does it start? Since a star is made of matter, evidently it must form from diffuse matter by a process of accretion. Is there matter scattered around in space? Plenty of it, as we have seen; gas and dust clouds are found all over our galaxy, and they are immense. Hence the material is there. Now we must see how this material coalesces into nuclei of condensation around which stars form. Let us consider the problem in general terms.

We know that the atoms that make up the interstellar medium are in motion. And we know also that the density of this interstellar medium is very low. For a nucleus of condensation to form, a number of atoms must pass so close to each other that each will be affected by the gravitational attraction of all the others and will not be able to continue on its way by the law of inertia. But the force of gravity is very weak—indeed, the weakest force known. The chance encounter of a few atoms cannot produce a nucleus of condensation because their mutual attraction is not strong enough to keep them together. But suppose chance, or some process we shall not investigate, brings together in a limited space such a large number of particles that gravity can take hold. Now the particles remain trapped together. This gives rise to a new galactic object, a mass of gas and dust, a protostar, invisible to all optical insturments. The idea of an invisible entity may be disconcerting, but there is no reason why invisible objects should not exist. I am not asking you to believe in ghosts, but only to accept the fact that there are things in our world—such as protostars, for example (or at least most of them)—that cannot be seen. Obviously, the chance that particles will meet in sufficient numbers to produce a protostar is not zero and is greater where there is a lot of gas and dust (namely, within the galactic disk, and particularly in the spiral arms).

You may remember the dark globules, those dark features that can be seen projected against bright nebulae. Observe figures 90 and 91 which are photographs of the nebula NGC 2237, or the Rosette nebula, in the constellation Unicorn. The two dark points indicated by arrows in figure 90 are probably protostars (but they are not the only ones). But then we *can* see them? Yes and no. Let us say that we can see them as we "see" the moon in broad daylight projected against the solar disk during a total solar eclipse. If the globules happened to be in a dark region of the nebula, they would not be seen. How many globules are there? The Rosette nebula, for one, appears to be a veritable crucible of stars. The great nebula in Orion is likely to be another hotbed of stars, particularly in view of its high density. But there are many other nebulae that might be, since they satisfy all requirements; hence the number of globules must be large.

You may wonder how large an aggregate of particles must be in order to constitute a protostar. Let us speak plainly. How many atoms must come together to form a protostar? There is no simple answer. It depends on temperature. In a high-velocity field the formation of a protostar is evidently more difficult than in a low-velocity field because gravity has less chance to take hold among fast-moving particles. The speed of the particles increases with the temperature. Hence a larger number of atoms must come

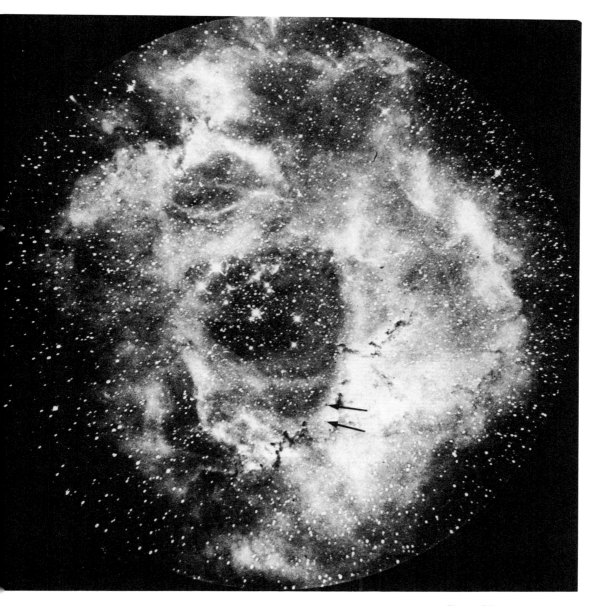

Figure 90
The Rosette nebula (NGC
2237). Arrows indicate
probable protostars.

Figure 91
Detail of the Rosette nebula
in the vicinity of the two
dark globules indicated by
arrows in figure 90.

together in a hot gas than in a cool gas. An estimate can be made for average conditions, however; it tells us that the number of atoms needed is 10^{56} to 10^{57}. It is a very large number, almost inconceivable, and we shall not even try to figure it out. All of the earth's nucleons (protons and neutrons) only add up to a millionth of 10^{57}. It would take all the nucleons of a million earths to make up the mass we need. But a million earths is equivalent to one sun. Thus the median number of atoms needed to form a protostar under average conditions is equivalent to the number of atoms in the sun; in other words, the mass we find is on the order of one solar mass, that is, the average stellar mass, which is just right, just what we would have liked to find.

Although it is very gratifying to see one's wishes come true, we cannot stop here. Once started on a certain path we must go on until every phenomenon is explained, every observational datum falls into place, and every theoretical deduction finds a counterpart in the sky, or at least some observational confirmation.

•FROM PROTOSTAR TO STAR

It has been some time since we left our protostar. If we go back now, we shall find that things have changed. The protostar has become smaller and denser. What has happened, and is still happening, and will continue to happen for a while longer, is that the original condensation has collapsed under the pull of gravity. The atoms have fallen inward, just as a stone falls toward the center of the earth. It takes time for the atoms to finish falling because the protostar is a vast mass of gas, on the order of 1 light-year across. In the absence of obstacles a falling object will fall faster and faster (that is, will accelerate). A gas whose particles move increasingly faster is a gas that gets hotter. Once this process begins, certain consequences are inevitable. The temperature, which originally was roughly 100°K (about −170°C), increases.[68] When it reaches 50,000°K in the central region, the average velocity of the atoms has increased from a couple of kilometers per second to about 40 km/sec and the volume has markedly decreased; the diameter of the original condensation is reduced to about 150×10^6 km, which is more or less the radius of the earth's orbit. At this point the hydrogen and helium atoms in the core, the densest and hottest part, move fast enough to tear away (by collisions) the electrons from around each others' nuclei. The gas in the stellar core is now essentially a mixture of electrons and nuclei of hydrogen and helium.

The protostar is no longer a dark object; like any other heated body, it radiates according to its temperature. When the surface temperature is high enough for the emitted radiation to contain photons of the visible wavelengths, it becomes visible. It could have been seen earlier, had we made observations in the infrared, because a body's temperature need not be very high for it to radiate at these wavelengths. The protostar appears now as a faint, fuzzy object that can be assigned a point in the H-R diagram. As the protostar evolves we can follow its transformations and mark different points on the diagram. Of course, it cannot be done quite this way. Nobody lives long enough to observe the changes that take place in a star. But it can be safely assumed that the stars did not form all together, from which it follows that the observed protostars are in different stages of evolution. If we put all the known protostars in the diagram, the tracks we draw represent the evolutionary tracks of average protostars of different masses.

Unfortunately, protostars are difficult to observe, or rather, they are difficult to recognize as such. There is good evidence, however, that T Tauri variables, among others, are actually protostars. They are G and K stars that fluctuate irregularly in brightness and are found in very young clusters. In the H-R diagram for the cluster they belong to, they lie to the right of the main sequence. Observations show that most of the radiation emitted by these stars falls in the far-infrared. The explanation for this is that the star is immersed in a cloud of dust (the very dust from which it formed) that absorbs the radiation from the star and reemits it at the longer wavelengths corresponding to its own lower temperature, that is, in the infrared. In any case, protostars are objects difficult to observe. But their evolutionary tracks can be calculated theoretically by selecting various initial masses, and then deducing the consequences necessitated by the laws of nature.

To begin with, the protostar will continue to contract. Accordingly, the velocity of its particles will continue to increase, and the gas will become hotter and hotter. When the diameter is about 100×10^6 km, the internal temperature is 150,000°K and the surface temperature 3,500°K. Seen through the telescope, the protostar is now a blurred globular object. Although its surface temperature, and hence its emissive power (flux/cm^2), are not very high, the star is very large; with a diameter 65 times greater than the sun's, its radiating surface is 4,000 times greater than the sun's. This star-that-is-not-yet-a-star is nevertheless a conspicuous object; it can be observed and studied. Owing to its low temperature it is red in color; hence it falls on the right-hand side of the H-R diagram. Since it radiates a huge amount of energy, it is very bright; hence it lies high in the diagram. This means that it is in the region of the red giants. It is not a red giant, however; a red giant is something else, a bona fide star, an old star that has arrived in this region from other parts of the diagram, whereas the protostar is showing up here for the first time. It is here that the evolution of the star begins to be tracked, and it is here that the star will return, if its mass is large enough, toward the end of its existence.

Theoretical calculations show that at this stage the protostar is in rapid evolution; consequently, its position in the H-R diagram changes quickly. This evidently makes the observation of protostars even more difficult. The region of the diagram where they should be represented is in fact sparsely populated. But this is not unexpected; since stars spend but a short time in this region, it follows that at any given time very few are found there. Conversely, the heavily populated regions of the diagram correspond to conditions of stellar stability. This points to another important aspect of the H-R diagram; it indicates the conditions in which a star finds itself for a good deal of its life as a stable system.

Figure 92 shows the evolutionary path of a star of one solar mass. It is the very path our sun must have traveled to become the star it is today. The first stage (corresponding to point 1) takes about a thousand years; it starts more or less at the time of condensation. Subsequently, the star contracts fairly rapidly. There is a marked decrease in luminosity because the increase in surface temperature is not enough to compensate for the shrinking of the radiating surface. Ten million years from its formation the protostar is a little larger than the sun (point 2). The internal temperature has risen vertiginously; it is now on the order of 10×10^{6}°K. At this point the protostar becomes a star.

At a temperature of 10×10^{6}°K the atoms (or, more precisely, the atomic nuclei) have enough speed to overcome the electrostatic forces that tend to keep them apart. Beyond a certain point the powerful forces of nuclear attraction come into play and fusion occurs. The star's core ignites and starts producing energy by "burning" hydrogen into helium. The temperature rises. For a while there are variations in size and luminosity, but finally the

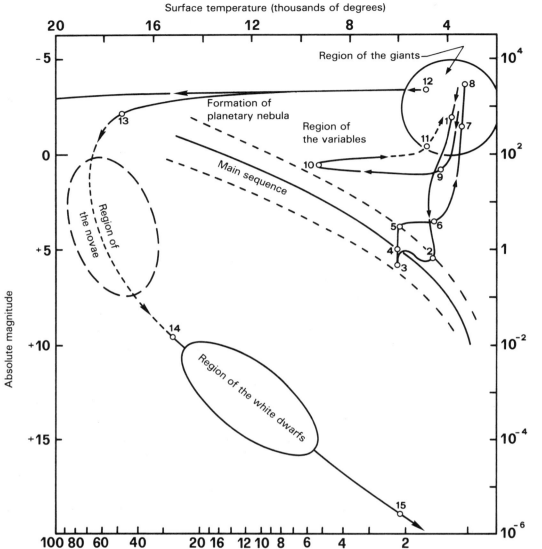

Surface temperature (thousands of degrees)

Region of the giants

Formation of planetary nebula

Region of the variables

Main sequence

Region of the novae

Absolute magnitude

Region of the white dwarfs

Surface temperature (thousands of degrees)

Luminosity (sun = 1)

Figure 92
Evolutionary path (points 1, 2, . . . , 15) of a star of one solar mass in the H-R diagram. The region of the novae is also indicated. The temperature scale at the top applies to points 1 to 12 of the path. After point 12, however, the path bends downward, and points 13 to 15 on it now are with reference to the temperature scale at the bottom.

star settles down; it now has a diameter slightly smaller than that of the sun and a surface temperature of about 6,000°K. Twenty-seven million years after it formed (point 3) the star has found a stable place on the main sequence. It will remain on it for 10 billion years.

At this point our star (a star of one solar mass—the sun, if you like) is at the so-called zero age. It has stopped contracting because the force of gravity, which would tend to make it contract even more, is at last balanced by the forces that tend to make it expand (gas and radiation pressure). As long as helium is produced from the fusion of hydrogen nuclei, the star does not change, or rather, it does change, but very slowly. Evidently, as more and more hydrogen is burned into helium the percentage of hydrogen in the core decreases, while the percentage of helium increases. The next step takes place within the main sequence, as shown in figure 92. Little by little the star becomes brighter, and 4.5 billion years from the zero age it attains the sun's luminosity (point 4). It has slightly increased in size; its diameter is now 1.5×10^6 km, which is the sun's diameter. The surface temperature, and hence the spectral type, has not changed. The increase in luminosity is due to the increase in size.

As far as the sun is concerned, we could stop here. Theory tells us how it has evolved thus far. To have reached the point in the diagram where it now lies, the sun must have formed from an interstellar cloud about 4.5 billion years ago. (Let us ignore the 27 million years spent in the protostar stage, which are nothing compared with 4.5 billion years.) During all this time nuclear transformations have been going on without cease in the sun's core. Hydrogen nuclei have been combining to form helium nuclei, and in the process energy has been released. What is happening in the core makes the sun a stable, self-regulating machine—one might say a perfect machine. If for some reason energy production were to decline, there would no longer be hydrostatic equilibrium, and the overlying masses would start to collapse. As a result, the central temperature and pressure would rise, causing an increase in energy production that would not just halt the contraction but bring things back to the way they were. If, conversely, energy output were to decline, the sun would expand. The gas pressure on the iner layers would decline, the temperature would fall, energy production would decrease, the expansion would stop, and everything would go back to the initial state. These feedback mechanisms counteract any small change that may occur, so that the machine can go on working, self-regulated, for billions of years.

•FROM THE MAIN SEQUENCE TO THE REGION OF THE RED GIANTS

Theory has told us what happened to the sun in the past. What will happen to it in the future? Let us follow the sun's evolution without pretending to be too precise about it.

Hydrogen burning is the determining factor in the history of a star like the sun; it dominates about 90% of its life cycle. Long as it may be, this period must come to an end when the core of the star becomes helium. Till then there is no renewal of material in the core; so long as hydrostatic equilibrium prevails, the hydrogen surrounding the core cannot fall inward and remains "suspended," so to speak. In other words, as long as hydrogen burns into helium, there is a central furnace, about one-tenth the size of the star, that produces energy and supports all of the overlying mass. What happens when all the hydrogen in the core is exhausted and the latter consists of helium alone? The production of energy stops. No longer supported by the mechanism that has worked till now, the core starts to collapse under the pull of gravitation. When the star is 9.2 billion years old, it begins to depart from the main sequence, moving to point 5 in figure 92. At this point the star is about 1.5 times brighter than the sun and has a diameter of about 2 million km.

The fact that the helium is falling inward simply means that the gas is in rapid contraction. The temperature rises, and this in turn causes the hydrogen surrounding the core to become hotter and to start burning into helium, as happened previously in the core. The internal structure of the star has changed. There is an inert helium core, a hydrogen shell burning into helium, and a second hydrogen shell that does not burn. As the core keeps shrinking, the temperature keeps rising, and energy production in the first hydrogen shell increases. But instead of becoming more luminous, the star becomes larger because the increase in energy output serves to raise the temperature in the second hydrogen shell, though not to the point of igniting nuclear reactions there. Thus the energy produced goes into inflating the star. This expansion in turn causes a decline in surface temperature, and the star becomes increasingly redder. In a little more than 1 billion years, a tenth of the time spent on the main sequence, the surface temperature falls to about 4,000°K. Because the expansion and the decline in surface temperature are concurrent, the star's luminosity remains the same. The core continues to contract, however, and the production of helium continues to increase. The energy output becomes hundreds of times greater than it was when the star was in the main sequence. Part of this energy can no longer be absorbed by the second shell and escapes into space. Thus in this phase the luminosity increases along with the size, and in about 100 million years the star becomes a red giant.

In figure 92 the path from point 5 to point 6 marks the period of constant luminosity and for obvious reasons is parallel to the horizontal axes. At point 5 the star has doubled in size from what it had been at the age zero. At point 6 it has again doubled in size and has a diameter of 3 million km; but the helium core is very small, with a diameter from 0.001 to 0.01 that of the star. Figure 93 shows the variations in size during a star's evolution.

The last piece of the evolutionary path is very rapid. The surface temperature declines further, but the luminosity increases tremendously. Obviously, it is the diameter that has increased. It is now 150 million km, which means that the star would occupy a half of the earth's orbit. Although it contains a quarter of the star's total mass, the helium core continues to be very small; the first hydrogen shell is also very thin—only a few thousand kilometers. The rest of the star is a vast hydrogen cloud of very low density—2×10^{-8} g/cm^3, on the average (approximately 10^{16} hydrogen atoms/cm^3), or about 0.00002 the density of the air we breathe.

The helium core keeps on contracting a while longer, and the star expands until it becomes a true red giant (point 7 in figure 92). When the central temperature reaches 100×10^6°K, the helium nuclei combine to produce carbon nuclei. The energy released in the process raises the temperature of the core. One might expect the core to expand until radiation pressure and the force of gravity balance each other, but this does not happen. Theory says that the core cannot expand much, owing to the very high density of free electrons. As a result, the core cannot cool down fast enough; this accelerates the fusion of helium nuclei, which in turn causes a new rise in temperature, and so on. In a few hours (almost instantly, so to speak) the helium core becomes so hot that it explodes. This is the phase of the so-called "helium flash" (point 8). Torn by the explosion, the core expands at great speed and

Lines of equal radius in the H-R diagram. All the stars located on any of the parallel lines have the same radius. In the course of its evolution a star moves to different parts of the diagram and consequently changes its size; it expands when it moves to the right, and it contracts when it moves to the left. According to the relation $L = 4\pi r^2 \sigma T_e^4$ (see the simple model of the stellar atmosphere in part II), for these lines L/T_e^4 is a constant. $R_\odot$ is the radius of the sun.

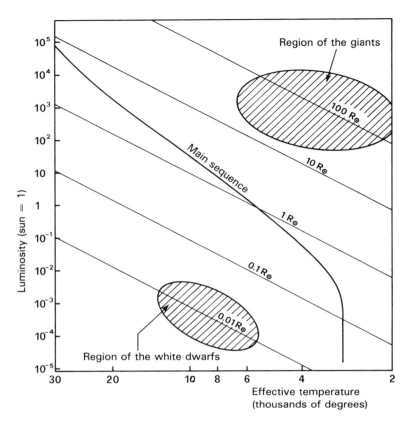

the internal structure of the star changes rapidly. The temperature falls markedly in both the helium core and the first hydrogen shell and the production of carbon nuclei stops. With less energy emitted, the luminosity declines abruptly. No longer supported, the gaseous mass begins to collapse and the star shrinks.

In this new phase of contraction (point 8 to point 9), the helium core is again compressed, with a consequent rise in temperature. Eventually the helium starts burning into carbon again. Ten thousand years after the flash the central temperature is 200×10^6°K, and helium is the main producer of the energy emitted by the star. Slowly the nucleus expands while the gaseous envelope contracts. Although the diameter shrinks, the surface temperature rises, and this produces another phase of constant luminosity. The star moves from right to left in the H-R diagram; from point 9 to point 10 in figure 92. During this period a carbon core begins to form within the helium core, just as before a helium core had formed within the hydrogen core; it is the same story, with different elements.

At point 10 the structure of the star has become more complicated. At the center is a carbon core, which begins to contract and raise the temperature, thereby igniting nuclear reactions in the inner part of the helium shell. Then comes a hydrogen shell, whose temperature and density are high enough for nuclear reactions. And finally there is an enormous hydrogen envelope, which does not take part in the nuclear processes. The carbon core does not produce energy, but by contracting it keeps raising the temperature of the helium shell. Consequently, the overlying layers absorb energy and heat up, causing a new expansion of the star. The surface temperature declines, the luminosity increases, and the star returns to the region of the red giants (point 10 to point 11). It all happens in a few million years for a star such as the sun.

At this point the story becomes somewhat unclear. It seems likely that the star goes to the region of the red giants not directly but by traveling a zigzag course similar to the one it followed after the first flash, though increasingly tighter. Perhaps there are several flashes.

The region of the flashes and zigzag movements is occupied by the various kinds of variable stars. This is in agreement with the evolutionary models because starting from the first helium flash, the predicted path is very sensitive to small variations in chemical composition and physical conditions. The pulsations envisaged by the models correspond well, on the whole, to the pulsations observed in the Cepheids. More generally, theoretical predictions are in good agreement with observations here, even though further research is needed to clarify various details.

One might expect nuclear reactions to take place in which carbon nuclei combine to form atoms of heavier elements. But if the mass of the star is not large enough, this cannot happen. The rise in temperature in the central regions, due to the contraction of the core, is never so great as to ignite these reactions, which require at least $600 \times 10^{6\circ}$K. Instead, the rise in temperature causes an increase in the amount of energy produced by the helium shell surrounding the core, and the star, as we have seen, turns again into a red giant.

Apart from what happens to the star in this period of instability, it seems certain that soon afterward the star enters the final phase of its life cycle.

·FROM RED GIANT TO BLACK DWARF

As the star expands, the temperature of the upper layers falls to the point that ions can recombine with electrons to form neutral atoms. Thus the absorption and reemission of photons by neutral atoms begin to play a part by causing the upper layers to heat up and expand further. The expansion lowers the temperature, which leads to the formation of more neutral atoms that absorb energy. This results in a new expansion. And so on. The upper layers expand faster and faster and eventually escape into space, leaving behind a small white object—the core.

What we see now—a central star with a gaseous envelope of neutral atoms—is a planetary nebula. The star continues to work as before through the helium reactions. The total luminosity has remained unchanged, but the diameter has undergone drastic variations. Accordingly, the position of the star varies considerably in the H-R diagram (between points 12 and 13 in figure 92). The surface temperature has quickly become that of the outer part of the core (on the order of 10^4°K). Hence the diagram's path is horizontal, with movement from right to left. Not much is left of the original star—a contracting carbon core surrounded by a helium shell that produces more carbon. In time the free electrons become so dense that they can no longer be compressed, and the contraction stops. This happens when the star's diameter is reduced to about 30×10^3 km. The density of matter is now an unbelievable 10^6 to 10^7 g/cm³. The star has become a white dwarf.

The details of the star's evolution are not too clear from the time it loses the gaseous envelope (point 13) to the time it can be called a white dwarf. The theories of stellar evolution offer no help in tracking it through its "last hours" as an object shining by nuclear processes. Furthermore, there are scant observational data because only a few stars are found in the region between points 13 and 14. But the last stage of the star's life is well understood (starting from point 14), partly because of the large number of white dwarfs.

Although its surface temperature is fairly high, the star is a hundred times less bright than the sun and is definitely nearing the end. The high surface temperature is simply because between its inner and outer layers there are no screens (absorbing layers) as efficient as there used to be before the star became a dwarf. The radiation emitted by its central regions is not absorbed by the star because the star is now composed essentially of nuclei and electrons; that is, it lacks the atoms that can absorb and reemit radiation. Hence the surface temperature cannot be very different from that of the central regions. The difference is at most a factor 10—10,000°K, 20,000°K on the surface, 100,000°K at the center. A far cry from the 12×10^{6}°K that started the star's life cycle. We are really at the end.

In time the star shrinks some more and expends the rest of its energy. It can no longer produce energy—not by nuclear reactions, since the fuel is long since exhausted; nor by gravitational collapse, since its matter is too dense to be compressed any further. At this stage, therefore, the star radiates like any other heated body. As the temperature slowly falls, the color changes from white to yellow, then to red and deep red; gradually, the star blackens, like a burned-out cinder. The time necessary to complete this stage could be extremely long because as the temperature declines, so does the emission of radiation. But at a certain moment the core begins to solidify. By releasing energy this process accelerates the cooling of the star. When the star is completely solidified there is no more energy to be extracted, and the star is finished. According to theory, this last chapter in the star's history takes about a billion years. This means that we should be able to observe quite a number of white dwarfs. And in fact we do. Thus the white dwarfs seen today are what is left of main-sequence stars that have come to the end of their active existences in the last several hundred million years. If there are stars that ceased earlier, and there is no reason why there should not be, we have no hope of seeing them. They are black dwarfs—dark relics that travel through space undetected.

In all probability this will be the end of our sun. No point in getting sentimental about it. We have already agreed that there is nothing in the world that lasts forever. Five billion years from now, take or leave a billion, all the stars like the sun, or a little more massive, will have become black dwarfs.

·HISTORY OF A STAR MORE MASSIVE THAN THE SUN (AND, INCIDENTALLY, OF ONE LESS MASSIVE)

What happens to a star more massive than the sun? It is obvious that the theoreticians will try to apply their equations to all possible cases. They are not always successful, however, because the problem for the case of a massive star is more complicated than the one for the case of a star the mass of the sun, which travels its evolutionary path relatively undisturbed by other phenomena. But often things work out very well—for example, when *only* the mass, among the various parameters, is made to vary.

Let us see what happens to a star in every way just like the sun except that it is more massive. The first part of its life cycle is not very different, all in all. There is the stage of protostellar contraction, then the start of nuclear reactions that convert hydrogen to helium, and then the main-sequence stage. Unlike the sun, however, the star lies on the upper part of the sequence; the more massive it is, the higher up its position. This we knew from the mass-luminosity relation. But there is a very important fact to be considered. The production of energy increases with mass. Theory is quite clear on this point, which is also confirmed by observation. The increase in energy output might be expected to be proportional to the mass. It turns out, instead, that a massive star is more luminous than a less massive star by a greater factor than the ratio of their masses. A simple calculation will prove it. The sun emits a flux of 4×10^{33} ergs/sec and has a mass of 2×10^{33} g. The ratio of these two numbers tells us that the amount of energy produced by each gram of solar matter is, on the average, 2 ergs/sec. But there are stars 10^4 times brighter than the sun whose masses are only 20 times the sun's mass. It follows that each gram of their matter must produce, on the average, 500 times the energy produced by each gram of solar matter. Similarly, these are stars 10^4 times less bright than the sun whose masses are only one-tenth the sun's mass. Hence in these stars the production of energy by each gram of matter is, on the average, a thousand times smaller than in the sun. This means that massive stars burn up faster than lighter stars and therefore have shorter lives. The reason, as you may have guessed, is that massive stars become much hotter in the center, and the production of helium from hydrogen increases rapidly with the temperature. An idea of the increase is given by the following: If the temperature doubles, the production of helium increases by a factor of 3×10^4. It is evident therefore that the evolutionary processes speed up in massive stars—the more so, the larger the mass. The life of a star of 10 solar masses is 100 times shorter than that of a sun-type star, roughly 100×10^6 years. Should there be a planet around such a star, many things that have happened here, around the sun, would not happen there. Certainly there would not be enough time for a planetary evolution resulting in life forms.

On the other hand, stars less massive than the sun live longer—the longer, the smaller the mass. A star one-tenth the solar mass would live 100 times longer than a sun-type star, roughly 1,000 billion years. The evolution of these stars is pretty well understood; they burn up slowly because their hydrogen reactions are exceedingly slow. According to current views—on which we shall not elaborate in this book—the universe began more or less 10 billion years ago. If a star one-tenth the solar mass, or smaller, formed at the same time as the universe, it would not have come to an end yet. It would be burning slowly and shine feebly, practically without change all the while. Nearby planets would have had more than enough time to develop biological life—though not necessarily human, of course.

Let us get back to massive stars. At the top of the main sequence lie the so-called blue giants, which are very large, very massive, and very hot. These stars pass quickly through evolutionary stages similar to those I have described for the sun. Things change when the star has evolved to the stage of the inert carbon core surrounded by a helium shell that produces new carbon by nuclear reactions. The contraction of the core is such that long before the star would shrink to a white dwarf, the central temperature reaches the 600 million degrees necessary to start carbon reactions that result in the formation of magnesium and other heavy elements. At this point the star stops contracting, but only as long as the carbon lasts. Then it starts contracting again, with a consequent rise in temperature. What takes place, in brief, is a series of contractions alternating with periods of stability during which new nuclear reactions start that produce increasingly heavier elements. If the mass is large enough, the process goes on until iron atoms are formed.

This marks the end of the star. Atomic physics shows that iron atoms do not release energy while combining; on the contrary, they need energy to combine. Thus iron cannot be used as a fuel. This is why the accumulation of iron in the core brings nuclear reactions to an end. The star starts contracting again, but this time the contraction goes on to extreme consequences—to densities of more than 10^{12} to 10^{14} g/cm^3, pressures of 10^{26} atmospheres, and temperatures of 10^{10}°K—until the nuclei are so jammed together that there cannot be any further contraction. All of a sudden the star is rent by a mighty explosion that flings at least half its material into space. This phase—the last catastrophic contraction followed by an equally catastrophic explosion—lasts but a few minutes. And we have a supernova. The star becomes billions of times brighter, sometimes as bright as a whole galaxy. In the final instants temperatures are so high that many nuclei are torn apart, releasing protons and neutrons that combine with other nuclei to form heavier elements, up to uranium. When the star explodes, all these elements are injected into space, "seeding" it, as it were.

To summarize, the elements up to iron are manufactured within a massive star during the normal course of its evolution. The heavier elements, instead, form in the last moments of its life as a massive star. The dissemination of the elements occurs in the explosive phase. The time available for building elements heavier than iron is very short indeed; this is why the percentage of these elements in the universe is very low.

All that remains after the explosion of a massive star is a very dense core. Is it another stellar core on its way to becoming a white dwarf? Theory shows that the radius of a white dwarf depends on its mass and chemical composition, chiefly the former—the larger the mass, the smaller the radius. The typical radius of a white dwarf should be a little greater than that of the earth, and on this point observations are in agreement with theoretical predictions. Theory also shows that the radius is zero when the mass exceeds 1.2 to 1.4 solar masses. This means that a star more massive than 1.4 solar masses cannot be a white dwarf, since it would have to be infinitely small as a result of a total contraction. At the same time atomic physics shows that for a mass in excess of 1.4 solar masses the contraction results in such high densities that the negatively charged electrons penetrate the nuclei. By combining with protons, which carry a positive charge, they form neutrons, and ultimately we would have a star consisting almost entirely of neutrons, that is, a neutron gas so dense that it can support its own weight. The collapse stops because the star is now a stable system. It has become a neutron star—a superdense object of extremely small dimensions. To give you an idea of the density, a mass of the order of the solar mass would be contained in a sphere about 20 km across.

The existence of these objects, predicted by theory, has been confirmed by observations, and today nobody would dream of doubting it. Neutron stars, for example, have been observed in the middle of supernova remnants.

The pulsars are nothing but neutron stars. Since their discovery in 1967 about a hundred such objects have been found. All but two are exclusively radio sources, in the sense that only two of them coincide with stars observed optically. All the others are points in the sky that emit radio waves in short bursts at very short intervals. The energy distribution of their radio emission shows that it cannot be thermal radiation, that is, radiation emitted by hot matter simply because it is hot. It is explained instead as synchrotron radiation due to high-speed electrons moving in a magnetic field. Observations show that the pulsars are at distances ranging from 30 to 300 parsecs; hence they are galactic objects. They are not uniformly distributed in space, but tend to be concentrated on the galactic plane.

Let me expand a little on the interpretation of these pulsating stars. The extreme regularity of the pulses might suggest eclipses in a binary system, but this solution is unacceptable because the orbital velocities deduced for such a system turn out to be much too high. Furthermore, the two components should be moving rapidly apart; the orbital period, and hence the period of the pulsation, should grow longer much faster than observations show. Another possibility might be a single pulsating star. This solution has been discounted, however, for theoretical reasons. The accepted interpretation is that a pulsar is a rapidly spinning neutron star. In this view the emission originates from a particular region of the star; as the star rotates this region comes briefly into the line of sight, and we receive a short radio pulse. This effect is not unlike the flash of light from a lighthouse. A rotation that occurs in a fraction of a second—or at most 4 sec—implies an angular velocity so high that it would tear apart a large star; thus it must be concluded that the star is extremely small. Calculations show that a neutron star is just the right size. There is nothing strange about the fact that a star can rotate so fast. As every skater knows, you spin faster if you gather your arms in close to your body. This is a concrete example of a physical law known as the conservation of angular momentum. Accordingly, an extended stellar mass increases its rotational speed as it contracts and becomes concentrated. When it reaches the size of a neutron star, it is bound to be spinning very rapidly.

As far as the rotation is concerned, there is no problem. What is not so clear is the actual source of the pulses. Withoug going into the various models, let us just say that the lighthouse effect is the most satisfactory explanation. Contrary to what I said earlier, therefore, the pulses are not due to a pulsation of the star as a whole, but rather to the emission from particular regions. Given the very high rotational speed deduced from the shortness of the period, the pulsar must be an extremely small object—a neutron star.

If this model is correct, there should be quite a number of neutron stars in the sky, many more than are observed. First of all, having the "lighthouse" pointed precisely in our direction must be a rare case. Besides, one must take into account the dampening effect of interstellar clouds and the fact that a neutron star is probably a short-lived object.

If the model is correct, I said. What I mean is that although this model is the best that can be produced today, it still leaves various matters unexplained. It is not clear, for example, why there are pulsars whose periods are growing shorter instead of growing longer as they ought to. Nor is it clear why most pulsars lack optical counterparts, that is, neutron stars observable by telescope.

Go back a moment to figure 65. The two photographs clearly show the optical pulsation of a pulsar. Important as they are, photographs only show us a fact and at most enable us to describe it. But now we have an interpretation, though not entirely satisfactory. And this is what scientific research is all about: not just describing facts, but explaining them.

· THE UNCERTAIN END OF A STAR MORE MASSIVE THAN 3.2 SOLAR MASSES

Are all massive stars destined to become neutron stars? Theory shows that there is an upper limit to the mass of a neutron star of 3.2 solar masses and that a more massive star can never reach a state of equilibrium. To put it another way, no matter how much the star contracts, it never reaches a condition such that the central density can compensate for the weight of the overlying mass. Hence the contraction must continue until the star becomes infinitely small and infinitely dense. Shortly before reaching this limit, the star attains the size defined by the so-called Schwarzschild radius.[69] If the sun could shrink that much it would become an object 6 km across with a density $D = 2.5 \times 10^{16}\,\text{g/cm}^3$. Once it has reached the Schwarzschild limit, the star becomes what is known as a black hole, an object, that is, from which nothing, not even light, can escape and into which everything can disappear.

This is no joke. On the contrary, black holes are a very serious matter for people in our profession. I shall not say very much about them because to explain them properly I would have to introduce the concepts of general relativity, which are out of place in this book. All I can do is try to make you grasp the idea with simple words and familiar concepts.

First of all, let us take the theoreticians' word for it that there is no way to stop the collapse of a star more massive than 3.2 solar masses. If so, the force of gravity on the surface of the contracting star must increase indefinitely because while the mass remains the same, its distance from the center keeps on decreasing. At the same time the so-called escape velocity increases, that is, the velocity with which a body must must be ejected from the star in order to escape from the star's gravitational field. (At lower velocities, sooner or later the body would fall back.) If there is no limit to the gravitational pull on the surface of the star (which is contracting indefinitely), sooner or later the escape velocity will equal the speed of light. Beyond this point of contraction (the Schwarzschild radius) the velocity needed to escape is higher than the speed of light and therefore not even light can escape the star's gravitational field. Beyond the Schwarzschild radius there can only be a black hole.

Let me explain it another way. A consequence of the theory of relativity is that a beam of light must be deflected from its path in the vicinity of a mass. Solar eclipses have been used to verify this prediction. A star seen near the edge of the sun during a total eclipse should be found to occupy a slightly different position from the position it occupies when the sun is far from the line of sight because its light is deflected as it passes near the sun (figure 94). The effect predicted—and on the whole observed— for the sun's gravitational field near the edge of the sun is a displacement of 1.7″. Near the surface of a white dwarf it would be 1′, and 30° near a neutron star. Let us consider the latter. According to theory, the light that leaves a point of the neutron star's surface at an angle greater than 60″ from the normal to the surface will be bent back onto the star. Conversely, the only light that leaves the star is light that propagates at an angle smaller than 60″ from the normal to the surface. As the star's gravitational field becomes increasingly stronger, the angle from the normal below which light may escape becomes smaller; that is, the light must leave at an angle closer and closer to the normal—or, in other words, the light must leave in an increasingly narrower cone around the normal. It turns out that when the star is within the Schwarzschild radius, the aperture of the cone is nil. Hence not a thing can escape from the star.

Figure 94
Effect of the sun's gravitational field on a beam of light. The diagram is not in scale: (a) is a view from the top; (b), the view from the earth. If the sun did not deflect beams of light in its vicinity, the beam of light AB from star A grazing the sun's edge at B would proceed on a straight line to C', so that star A would not be visible on the earth (at C). But star A *is* visible because passing by the sun, the light beam is bent by α and changes its course to BC. From the earth the star is seen along the line CBA', that is, at A'. Six months later, when the region of the sky occupied by the sun is at 180° from the region occupied during the eclipse, the star is visible at night and is seen at A, that is, in a different point of the sky. The angle α is about 1.7″.

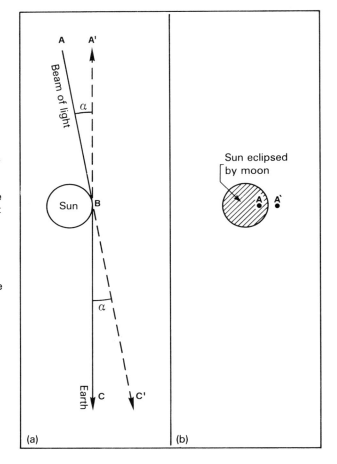

This explains roughly why the existence of black holes can be accepted, at least in principle. More correctly, I should say that their existence is a consequence of certain equations of general relativity. Should we assume that all the consequences of a theory correspond to reality? Certainly not. There are many examples in the history of science of theories that had to be abandoned because their consequences were contradicted by facts. The theory of relativity is one of the greatest achievements of the human mind, and there is no doubt that it explains many phenomena that would otherwise be unexplainable. There is no doubt, moreover, that many of its predictions do correspond to reality. But this does not mean that *everything* it predicts must necessarily happen. It has to be demonstrated, proved, verified. The fact that a concept is logically faultless is not enough for us to accept it without discussion. On the other hand, no concept that strains belief should be rejected out of hand; common sense is not always the wisest guide. At heart, for example, we are all Ptolemists; common sense says that the earth does not move. Even our language reflects it; we say the sun "rises" and "sets." But reason and observations have convinced us that it is the other way around. Try talking to simple people, however, and you will find that there are billions of Ptolemists. So let us not be slavish devotees of "common sense." Instead of rejecting the strange consequences of a theory, we should put them to the test. It is in the peculiar nature of the scientific method to abandon a theory if even one of its consequences is wholly contradicted by facts and to start looking for a better one. In our particular case I should add that there are theoreticians who either do not accept the black-hole solution or accept it as only one possibility. This is not absurd. If you have understood what a theory is, you should not be surprised by a difference of opinion. It is one thing to photograph an object and another to explain what it is. It is possible to make a model of a phenomenon, to find rules and laws that explain it; but those rules and laws may not exhaust all the possibilities, and their ultimate consequences may not find corresponding realizations. Then again, there may be a different model, different rules and laws, for explaining the phenomenon. All this is particularly true at the frontier of knowledge, as in this case. Finally, it should be kept in mind that a theory is not a discovery of an existing thing. It is an invention, a mental construct. A model is not necessarily reality. If it were, if theory did not depend on men who think and create new research tools, the history of science would be a series of forward steps without ever a change, a doubt, or a reversal of previous positions.

Let us return briefly to the black holes. As I was saying, the pull of gravity is so strong near of the surface of such a star that nothing, not even a photon, can escape. Although possessed of a considerable mass, it is invisible. It is known only by its gravitational effect on objects in its *immediate* vicinity. If the sun were to become a black hole, none of the planets would notice a thing (at least from the gravitational point of view). This is because the gravitational attraction of any homogeneous spherical object, such as the sun, on a second object *sufficiently far away* is calculated by assuming the mass of the former to be concentrated at a point (namely, its center, which is its center of mass). For all practical purposes a black hole is in fact a pointlike mass (figure 95). The trick in escaping its gravitational effect is to avoid getting too close to it.

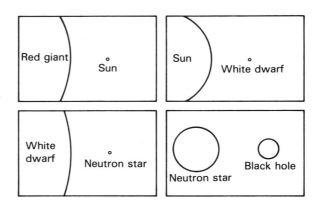

Figure 95
Comparing the sizes of different stars. Each figure is to a different scale (It would not have been possible to draw all the stars to the same scale in a single page-size figure.)

The black hole has other peculiar attributes, including the property of acting as a bridge between two adjacent universes or between different points of our universe—provided one approaches the hole with all due caution. This tells you that not all the characteristics of a black hole are bad. I shall not talk about them, however, for various reasons. In the first place, I do not know enough to find the right words, the simple language needed to explain very complex things without causing mental disorders. Furthermore, the whole matter is still at the level of theoretical speculation—speculation whose validity, limits, and significance are hard to assess. Finally, since the subject cannot (at least not yet) be explained in concrete terms, on the basis of concrete facts, I am afraid of confusing the reader who is not familiar with mathematical symbolism. Poorly handled, this topic is a prime example of the kind of thing that generates an irrational fear of the "mysteries of the Universe."

To close the subject, I shall say that some researchers believe they have found black holes, or at least some evidence of their existence. Let me explain. If the sun were a black hole (it does not have enough mass ever to become one), it would be invisible. But let us suppose that an outside observer still saw the earth and that observations showed it to be executing its customary planetary motions—in particular, an elliptical orbit. The observer would deduce that in the focus of the orbit is a mass of such and such a size that cannot be seen. He might conclude that the invisible object is a black hole of one solar mass. This is just an example, but there are actual cases where the observed effects could well be explained by black holes. Different explanations can be advanced, of course, but sometimes it is difficult to find many other possibilities; and all the available evidence, no matter how you look at it, seems to point in one direction. If the only possible solution is to postulate a black hole, then we should assume that that is what it is, until there is proof to the contrary. We have not come to this point, however, because there is no phenomenon that can only be explained by a black hole. Nevertheless, there are several approaches by which we can attempt to establish its existence.

Let us consider a specific case. Suppose you have a spectroscopic binary system but cannot find the companion (no spectrum, for example). Suppose, further, that the invisible companion turns out to have too large a mass to be a white dwarf or a neutron star. Obviously, a black hole would fill the bill. Such a binary is Cygnus X-1, which is an intense source of x rays. The observed star shows a variable radial velocity; hence it is assumed to have a companion. This massive, invisible companion appears to be the source of the x-ray emission. Cygnus X-1 is the best candidate for a black hole, but the matter is by no means settled. Some scientists have proposed other solutions—a multiple system, for example—based in part on the uncertainties and margins of error inherent in both the observations and the models of close binary systems.[70]

Another place where a black hole is suspected is a stellar system whose mass is much in excess of the observed luminosity. While black holes would not contribute to the luminosity, they would certainly add to the mass. On the other hand, the missing mass could be attributed to faint stars or even planets. The number of stars or planets needed to make up the mass can be calculated.

This line of reasoning can be applied to other cases. But for the time being, unless the theoreticians come up with a way of halting the seemingly irresistible contraction of a star more massive than 3.2 solar masses, we have to accept the theoretical possibility that black holes exist and that they originate from very massive stars.

·SOME FINAL REMARKS ON H-R DIAGRAMS

We should now take a final look at the various H-R diagrams in light of what we have learned. It is easy to see that the original diagram, which is for stars in the sun's neighborhood, is more complicated than the diagrams for either galactic or globular clusters. The reason is that the stars populating the sun's neighborhood originated in various ways at various times. In other words, the stellar population in this region of our galaxy is not homogeneous.

Star clusters present us with an entirely different situation, and their diagrams hold the key to the interpretation of the original diagram. The stellar population in a cluster cannot but be homogeneous because its components are stars whose masses vary but that formed at the same time from the same material, that is, stars of the same age and composition that are close together, not by chance, but because of their mutual gravitational attraction. As you know, while a cluster may lose some of its members, it cannot capture passing stars. Thus you can be sure that all the stars in a cluster have been there from the time of its formation, whether recent or far back in the past. The diagrams for clusters are fairly uncomplicated. Furthermore, while different from one another, they can be ordered logically, from diagrams that show only main-sequence stars, to diagrams that lack blue main-sequence stars but show red giants and supergiants, to diagrams whose main sequences are even shorter but whose regions of giants and supergiants are richer. The explanation of these differences is immediate if we accept the evolutionary theory I have described.

In fact, the H-R diagrams for clusters have actually helped to formulate the current evolutionary theory. It is no small wonder that everything should work out so well! Nevertheless, this theory explains most, possibly all, of the facts we have discussed. Nobody could ask more from a theory.

The fundamental point is that every cluster can be assigned an age. A young cluster will have a diagram consisting essentially of the main sequence (see the Pleiades, for example, in figure 83). The diagram for an older cluster will lack the upper part of the sequence because the more massive stars that should be there are also the stars that evolve faster; hence they have already left the main sequence to become giants or supergiants, whereas the less massive stars are still there. In other words, as a cluster grows older, its main sequence becomes shorter because more and more stars depart from it; at a certain point the number of red giants will decline because they have evolved and moved to other regions or disappeared altogether. The theory of stellar evolution, which gives the theoretical evolutionary paths for stars of different masses, allows us to estimate the age of a cluster from its "turn-off" point, that is, from the point at which its main sequence turns toward the region of the red giants.

In the oldest clusters the main sequence stops at absolute magnitude +4. On the basis of the theory, these clusters are estimated to be 10 to 15 billion years old and appear to be the most ancient objects in our galaxy. Consequently, this is believed to be the age of the galaxy itself.

It may seem strange that 15 billion years ago our galaxy did not exist, that the world of stars the ancients believed to be perfect, immutable, and eternal was not there at all. But the evidence is there. You might think that it is merely a fruit of theory, and I could not blame you. But let me point out two things. First, there is other evidence for the age of our galaxy (and the universe). Second, scientists do not make up things without a thought to the consequences and do not reach certain conclusions unless *forced* to do so by the evidence, sometimes overwhelming, provided by facts and their relations. That is why my main concern throughout this book has been to show you how astronomers work (except in this last part, where, unfortunately, I have had to give you the gist of things without too much in the way of explanations). I have tried to show that astronomy is a science built essentially on the observation of celestial objects and, whenever possible, the analysis of their radiations. From this point of view, astronomy is a most exciting undertaking. With little to start with one builds a whole world of ideas. Everything finds its place, every new piece fits with thousands of other pieces into the thousands of parts of the one overall picture. It is this self-consistency of the whole, this fitting of all the pieces into the wider scheme, that gives validity to scientific statements. Thus if you are told that from all evidence our galaxy is 10 to 15 billion years old, you can well believe it.

The galactic clusters show a wide range of ages, but the globular clusters are all very ancient objects. Their diagrams display a horizontal branch at absolute magnitude 0 and a red-giant branch whose components are brighter than the red giants in galactic clusters. For all globular clusters the turn-off point is at the same absolute magnitude, which means that they all have about the same age. But this is also the age of the oldest galactic clusters (such as M 67, for example), which, however, are richer in heavy elements. Thus this difference in chemical composition cannot be explained by age. The differences are slight,[71] since galactic objects consist essentially of hydrogen and helium together with a small percentage of heavy elements (from lithium on).

Recall now that globular-cluster stars are the prototypes of Population II, whereas the stars in galactic clusters are the prototypes of Population I. It is not just age that distinguishes the two populations because Population I includes stars of all ages. Rather, it is that Population II consists *solely* of ancient stars. No young stars rich in heavy elements have been found in globular clusters. Something must have happened in a distant past that prevented the formation of new stars in the regions occupied by Population II stars. This conclusion is in agreement with current ideas concerning the evolution of our stellar system.

Ten or fifteen billion years ago our galaxy was a vast gas cloud (consisting of hydrogen and helium) more or less spherical in shape and rotating slowly about its axis. It was certainly not a homogeneous cloud. Owing to the mutual gravitational attraction of its atoms, the galaxy underwent a process of contraction, but because the cloud was both dishomogeneous and rotating, the motion of the particles was not an orderly mass motion toward the center. This probably caused fluctuations and currents. Condensations formed within the vast gas cloud, and by the process described earlier these blobs of matter contracted into stars. The stars formed in groups (as they still do today in the nebulosities of the disk); these first stellar systems are the globular clusters. When huge amounts of gas condensed into star clusters, the protogalaxy was deprived of its "support," that is, the pressure that the gas exerted before condensing. Consequently, the sphere began to collapse and the diffuse gas fell inward to the plane of galactic rotation, concentrating there. The halo was purged of gas, leaving only the globular clusters to revolve about the galactic center of mass in orbits determined by Kepler's laws. The single stars seen in the halo today, also revolving in Keplerian orbits, are stars that formed singly, formed in smaller groups that soon broke apart, or escaped from clusters. The phase of collapse must have been over fairly quickly—perhaps a few hundred million years. Once the gas was gone, no new stars could form in the halo.

This model of galaxy formation explains three things: the fact that the globular clusters are about the same age and are all very old; the lack of young stars in the halo; and the difference in chemical composition between the two stellar populations. According to theory, the heavier elements are formed in the most massive stars and dispersed into space by supernova explosions. The ejected material changes the chemical composition of the interstellar medium, and from this new material second- and third-generation stars are eventually formed. The formation of new stars can only take place inside the disk because it is here that the primordial gas and dust are concentrated. And it is here that the new material ejected from the stars mixes with the old.

We have now come to the end of our voyage of discovery. The answers we sought to our questions, while neither definitive nor perfect in every detail, are on the whole quite satisfactory. The structure of a star and all its characteristics (luminosity, surface temperature, spectrum, radius, and hence position in the H-R diagram) are determined by three quantities—mass, initial chemical composition, and age.

The picture is coherent; it only needs to be refined. Unless . . . unless something new comes up that proves it all wrong and we have to start all over again. Honestly, the likelihood that this will happen is very small.

· CONCLUSION

In all probability, the stars formed and evolved in the way that I have described. Since the current models explain fairly well the observed phenomena, we can extrapolate into the future and say with some confidence that things will go according to theoretical predictions.

I wish to stress, however, that while the theory of stellar evolution has a good chance of corresponding to reality, it is not a certainty. Obviously, nobody can be really sure about what happened 10 to 15 billion years ago and what will happen 10 to 15 billion years from now. All we have to go on are observations made here and now. But we can count on a large array of observational instruments and all the theoretical tools that have been developed by physicists, mathematicians, and chemists; in addition, we have a good deal of imagination and resourcefulness. Armed with all these tools, generations of scientists have rebuilt our world piece by piece, trying to understand how it works. Based on their efforts, and assuming that the laws of nature are valid for all eternity (for 10 to 15 billion years, that is), we can say that if today the world works this way, then things must have gone such and such a way in the past and will go such and such a way in the future. Nothing is sure, but everyone agrees that at the present time, on the basis of our current knowledge, there is no better explanation. Many details are still obscure, and some problems are as yet unsolved, but the basic feeling is that our ideas are essentially correct and that we have understood how the world is made.

In this book I have touched briefly on some topics and left others entirely out, partly to keep the book within a reasonable size and partly not to complicate the picture. Nobody starts exploring a new town from the back alleys. First you look at the map and then, little by little, at all the rest, down to the sidewalks and out-of-the-way places. If you settle in that town, eventually you will come to know even the smallest pothole on your street.

My purpose has been to give a general idea of our galactic town, to show how celestial phenomena can be studied and understood. A world understood, that is, defined by laws and reasonable concepts, is no longer alien, forbidding, or hostile; despite its remoteness, it is a human, more friendly place. It has also been my hope to demonstrate that there are no mysteries— only problems. Easy, difficult, tremendous, formidable, anything you like, but only problems. The fact that we do not know, and may never know, how to solve them does not change a thing. An unsolvable problem is not a mystery; it is just a problem that cannot be solved. Practically speaking, unsolvable problems and mysteries may be the same thing; but psychologically they are quite different. Confronted with a mystery, people stutter, quake, fall to their knees, build golden idols. Confronted with an un-solvable problem, at worst people give up trying, accept their limitations, and turn to more constructive activities. They do not need to invent all-powerful gods and end up feeling like powerless pawns in a game whose rules they will never know. They can keep their humanity.

APPENDIX: UNITS

These prefixes are standard in forming unit abbreviations:

Prefix	Example	Equivalence
n (nano)	nm (nanometer)	10^{-9} (one-billionth) m
μ (micro)	μm (micrometer, micron)	10^{-6} (one-millionth) m
m (milli)	mm (millimeter)	10^{-3} (one-thousandth) m
c (centi)	cm (centimeter)	10^{-2} (one-hundredth) m
d (deci)	dm (decimeter)	10^{-1} (one-tenth) m
da (deka)	dam (dekameter)	10^{1} (ten) m
h (hecto)	hm (hectometer)	10^{2} (one hundred) m
k (kilo)	km (kilometer)	10^{3} (one thousand) m
M (mega)	Mm (megameter)	10^{6} (one million) m
G (giga)	Gm (gigameter)	10^{9} (one billion) m
T (tera)	Tm (terameter)	10^{12} (one trillion) m

The (unprefixed) units used in this book are, in alphabetical order, the following:

Unit (unit abbreviation)	Equivalence
Angstrom (Å)	10^{-8} cm
Centigrade (°C)	$(9/5)°C + 32 = °F$ ($0°C = 32°F$, $100°C = 212°F$)
Dyne (dyne)	Force required to accelerate 1 g at the rate of 1 cm/sec^2
Erg (erg)	Work (= force × distance) done by a force of 1 dyne acting through a distance of 1 cm
Gram (g)	0.035 ounce (approximately)
Hertz	Frequency
Kelvin (°K)	$0°K = -273°C$
Light-year (light-year)	Distance (km) that light travels in 1 year; 1 light-year = 300,000 km/sec (speed of light) × 31,500,000 seconds (seconds in 1 year) = 9.45×10^{12} km (approximately)
Meter (m)	39.37 inches (approximately)
Minute (')	1°/60 (angle measurement)
Parsec (pc)	Parallax of 1"; 1 pc = 3.26 light-years (approximately)
Radian (rad)	The number of degrees on the arc of length 1 on the circumference of a circle of radius 1; 1 rad = 360°/2 = 57.29+°
Second (")	$1°/(60 \times 60) = 1°/3{,}600$ (angle measurement)
Watt (W)	Power (= rate of work) = 10^7 erg/sec

·NOTES

1. See the appendix for unit abbreviations.

2. The great astronomer J. Kepler found that the ratio a^3/P^2, where a is a planet's average distance from the sun and P is its period of revolution around the sun, is the same for all the planets. In actuality, this is not exactly true, but for our purposes we do not need to be absolutely precise.

3. Forgive me for reminding you that 10^{27} means the number 1 followed by 27 zeros.

4. If we take into account the mass of the sun, we obtain a density 1,000 times greater, but it is still only 3 billionths of a gram per cubic decimeter ($0.000000003\ 3 \times 10^{-9}$ g/dm^3).

5. I feel duty bound to add that the smallest ones are a few kilometers across, whereas Ceres, the largest one, has a radius of 325 km. They may all originate from an exploded planet.

6. I have used words instead of writing the number 2×10^{30} because in some cases words are much more eloquent than dry numbers. It is only an underhanded psychological ploy, because in fact 2×10^{30} is exactly the same as 2 thousand billion billion billion. This is the first and last time I alert you to this kind of thing; in the future you will have to learn to pay attention to what people say.

7. Careful with the word "enormous"; recall what I said about big and small.

8. This is in accordance with Einstein's celebrated equation $E = mc^2$, which gives the amount of energy E obtained from the annihilation of a mass m of matter (c is the speed of light). In the matter of units, 1 erg = 1 dyne/cm; see the appendix.

9. A gas expands even if it is not heated, but heating it certainly favors the expansion.

10. Given the earth's distance from the sun, as many protons pass through 1 cm^2 every second as are contained in a cylinder whose base is 1 cm^2 and whose height is 450 km (velocity of the solar wind). There are 5 protons in 1 cm^3, each with a mass of 1.67×10^{-24} g. Since the same applies, more or less, for all the square centimeters located on the surface of a sphere having the earth-sun distance as radius, and since the number of seconds in a year is approximately 31.5 million, a few calculations will give us our figure.

11. Take a quarter and hold it at such a distance from your eye that it will cover the sun. The ratio of the size of the coin to its distance from your eye gives the angle in radians sought—about half a degree. Note: 1 radian is defined as the number of degrees in the arc of length 1 on the circumference of a circle whose radius is 1, so that 1 radian = 360°/ (circumference of circle of radius 1) = $360°/2\pi$ = 57.29+°.

12. °K means degrees kelvin; this scale gives temperatures starting from absolute zero, which is equivalent to -273 on the centigrade scale (°C). See the appendix.

13. Spectral lines and the Hα line in particular will be discussed in part II.

14. This is why the following dialogue does not mean the same to everybody: "Do you love me?" "Yes, very much." "How much?" "Very, very much."

15. The law of universal gravitation states that any two bodies attract each other with a force that is proportional to the product of their individual masses and inversely proportional to the square of the distance between them.

16. Wind velocities of 12 km/hour have been measured—a light breeze. But given the high density of the lower Venusian atmosphere, a mass of gas moving at this velocity does not make for a gentle caress; to take an extreme case, a slowly advancing wall can hit you harder than a fast-moving feather.

17. What a spectrograph is and how it works will be explained briefly later on.

18. This is the way I see it, but obviously the question is immaterial because my personal wishes count for nothing, for less than nothing, among the powerful forces that determine history.

19. Actually, how can anyone be sure not to have preconceived notions? Some things seem so evident, so obvious, so natural that nobody would take them for prejudices. Furthermore, the willingness to accept, against common sense, conclusions imposed by experience or reasoning might itself be a prejudice. It is true that it has worked well many times, but why should it *always* work? Things being as they are, we can only do the best we can.

20. The double prime means seconds of arc; a second of arc is 1/3,600 degree (see the notes to table 1). An idea of the smallness of a second of arc can be gained by drawing an angle of 1° with the aid of a goniometer, an angle-measuring instrument, and then breaking this angle up into 3,600 equal subangles.

21. Within 20 parsecs there are only 700 such stars.

22. Contradicting myself immediately, I shall not give errors for each measurement, but I set this down to the nature of the book, which is not intended as a treatise.

23. According to this relation, the ratio a^3/P^2 is the same for all the planets; see note 2 for the definitions of a and P.

24. The earth-sun distance is an astronomical unit, often abbreviated A.U.

25. I have already said this, but considering what is happening in the world, I feel the need to repeat it.

26. One of these stars, arbitrarily chosen, will be assigned magnitude zero. Obviously, it does not matter where we set the zero on the magnitude scale, just as it does not matter where we set the zero on the thermometric scale. For example, 0 on a centigrade thermometer (0°C) corresponds to 32 on a Fahrenheit thermometer (32°F).

27. log is the abbreviation for logarithm. In this book it is always to the base 10.

28. *Luminosity*, or intrinsic luminosity, is the energy emitted by a star per second. It is an intrinsic property of the star; that is, it does not depend on the observer. In the cgs (centimeter-gram-second) system, luminosity is measured in erg/sec (energy per second); in astronomy, it is measured in absolute magnitude (M). *Brightness* is the flux of radiation received by the observer, flux being the radiant energy received in unit time by a given surface area (a square centimeter in our case) at right angles. In the cgs system, brightness is measured in erg/sec/cm²; in astronomy it is measured in apparent magnitude (m).

29. Å stands for angstrom, which is a unit of length; 1 Å is equivalent to 10^{-8} cm; hence 4,000 Å = 0.4×10^{-4} cm = 0.4×10^{-3} mm, that is, 0.4 μm (micrometer, also called a micron and written μ).

30. There are formulas to express this:

$B_{\lambda_m} \propto T^5$,

where λ_m is the wavelength corresponding to the maximum emission B_{λ_m}, $\propto$ is the symbol for proportionality, and T is the temperature;

$\lambda_m T$ = constant = 0.288   (Wien's law);

$W = \sigma T^4$ (Stefan's law),

where W is the emissive power and σ is a constant (=5.672×10^{-5}). The values of the constants are tied to the centimeter-gram-second (cgs) system of units. Note that the emission corresponding to the maximum of the curve increases as the fifth power of the temperature and that the emissive power increases as the fourth power of the temperature.

31. Since $\kappa\lambda$ represents the absorbed fraction of the incident energy, when the incident energy is totally absorbed, $\kappa\lambda$ is obviously equal to 1.

32. Hence the emission coefficient for the blackbody is $B(\lambda, T)$. For this reason cavity radiation is also called blackbody radiation.

33. 10^{11} particles/cm³ is very low density compared, for example, with the density of the earth's atmosphere at sea level—less than one hundred-millionth of it.

34. It is straightforward to estimate from the weight of a mound the number of bread crumbs in that mound.

35. Of course, the theory of stellar atmospheres did not stop at this first rough approximation. The current models are quite sophisticated and account for a great many facts; but problems still remain for at least the next generations.

36. In first approximation. This is the last time I shall remind you because I do not wish to seem pedantic. But bear it in mind.

37. It is obtained from the measured energy by applying the necessary corrections for atmospheric absorption and other factors that affect the measurements and by taking into account the star's distance.

38. Let me repeat once again that they are not really temperatures because one can only speak of temperature in the case of thermal equilibrium. They are parameters. The color temperature gives an indication of the spectral distribution of the energy and the effective temperature gives an indication of the energy flux. They tell us the temperature of the blackbody of that color or that emissive power, and since in first approximation the stars are blackbodies, and so on and so forth.

39. And to think that in our school days we used to say that such things as tangents and radians were totally useless!

40. What counts here is the radial velocity, the rate of motion along the line of sight. The theory of relativity predicts a Doppler effect in the case of transversal motion (motion perpendicular to the line of sight) as well, but the effect would only be appreciable at velocities close to the speed of light; in practice, it is negligible.

41. Evidently, in the case of faint stars we cannot afford the luxury of high-dispersion spectra. For these stars dispersions of 200–400 Å/mm may have to be used.

42. In astrophysics all the elements except hydrogen and helium are called "metals." It is just jargon. In effect, hydrogen and helium are by far the most abundant elements in the universe; the others constitute impurities, so to speak, and are considered all together. One could say "the other elements," but by now it has become a habit and it is harmless.

43. Hertzsprung used color, but, as you know, color and spectrum are both indicators of the same parameter—temperature—and therefore give the same results.

44. Do you remember what I said about the meaning of adjectives like big, small, ugly, beautiful, and so forth? For the time being, calling the sun a dwarf star is only a question of labels. Clearly this adjective will have to be justified.

45. Recall the relation, given in the treatment of stellar magnitudes in part II, that ties the distance to the absolute and apparent magnitudes: $M = m + 5 + 5 \log p$.

46. On the definitions of 3' and 1", see the note to table 1.

47. There are only 330 stars whose proper motions are about 1". On the average, the stars visible to the unaided eye have a proper motion of 0.1".

48. The figure is misleading because the proportions are wrong. In reality the lines from O to S and S' are almost parallel, and therefore the angle formed by S"S' and S'S‴ is practically 90°.

49. $v_t = d \times \mu = \mu/p$ if μ (proper motion) is expressed in radians and p (parallax) in seconds of arc (which means that d is in parsecs). Since μ is the annual proper motion, we must divide it by 3.16×10^7, the number of seconds (of time) in a year. We must also divide it by 206,265 because μ is given in seconds of arc, whereas it must be expressed in radians (see note 11). Thus we have to divide μ/p by $(3.16 \times 10^7) \times (2.06 \times 10^5)$, or about 6.25×10^{12}. This done, we have v_t in parsecs/ sec. If we want to express it in km/sec, we have to multiply the last result by 3.09×10^{13}, which is the number of kilometers in 1 parsec. In other words, to obtain v_t in km/sec we have to multiply μ/p (with μ and p in seconds of arc) by $(3.09 \times 10^{13})/(6.52 \times 10^{12}) = 4.75$.

50. For the meaning of Populations I and II, see the discussion of stellar populations later in this part.

51. Type I supernovae belong to Population II and type II supernovae belong to Population I.

52. The term "instantaneous" in itself does not mean anything. Instantaneous with respect to what? A person's life span can be considered instantaneous with respect to a geological era. In our case instantaneous obviously means much shorter than a second.

53. Clusters should not be confused with constellations, which exist only in our fantasy and are arbitrary groupings of stars having nothing to do with one another that are seen in the same region of the sky only because of projection effects.

54. The difference between a star's observed color index and its color index when there is no interstellar absorption is called color excess.

55. The scattering power of dust is inversely proportional to the wavelength of the incident light, whereas the scattering power of gas varies inversely as the fourth power of the wavelength.

56. About $10^3 - 10^4$ atoms/cm^3. This is still a very low density, however, in comparison with the density of air.

57. I prefer "formation" to "birth." It may seem one of the many idiosyncrasies that people have, and perhaps it is, but I feel that talking of the birth and death of a star may generate misunderstandings, besides being needlessly anthropomorphic.

58. Assuming, of course, that the universe had a beginning.

59. I resist the temptation to tell you to figure it out by yourself as an exercise.

60. Forgive the monotony of "as we shall see later on" and similar expressions. I do it to call your attention to the subject, so that *later on* you will remember to link the facts to the explanations.

61. In the case of globular clusters the effect of space reddening must be taken into account. To this end one must construct a piece of main sequence with Population II stars that are not reddened or whose reddening can be estimated. For this purpose one uses high-velocity stars—

Population II stars in the vicinity of the sun (see the following, a final look at our galaxy).

62. Am I merely spouting rubbish when I point out that in the face of knowledge, superstitions crumble, with all the infamous consequences?

63. The angular velocity of one body rotating about a second body is the first body's velocity of rotation (that is, the velocity with which it moves in its rotation about the second body) divided by its distance from the second body. According to this definition, a point at the edge of a record playing on a turntable has a greater velocity of rotation than a point near the record's center (since both points must make a single rotation in the same amount of time, but the outer point has farther to go), but their angular velocities are equal.

64. For example, in a meeting with officials of the Ministry of Education, my effective cross section is much larger than that of a high school student; on the other hand, my effective cross section is much smaller—let us say zero—when it comes to sentimental encounters with sixteen-year-old girls. All this is sad to contemplate, but is perfectly reasonable from the point of view of nature and society.

65. Stellar matter is in hydrostatic equilibrium when the forces tending to push it toward the center of the star are balanced by the forces tending push it away from it.

66. Since the hydrogen atom is four times lighter than the helium atom, and since the mass of hydrogen in the sun is about twice the mass of helium, it follows that for each helium atom there are (about) eight hydrogen atoms.

67. Conditions of instability come into play; they are due to, among other things, radiation pressure, which opposes the contraction, and hence the formation, of the star.

68. The kinetic energy acquired by the atoms comes from the corresponding decrease in the potential energy they have in the gravitational field of the protostar. We can thus say that the protostar heats up and releases energy at the expense of its gravitational field.

69. It is given by the relation $r_s = 2GM/c^2$, where G is the constant of gravitation, M the mass of the star, and c the speed of light.

70. The component stars of a close binary system are distorted by tidal effects and hence are no longer spherical. Furthermore, they exchange matter, which flows from one star to the other, thereby creating disks and jets of material. All this complicates the dynamics of the system and introduces a margin of arbitrariness in the theoretical interpretation of the observational data.

71. The abundance of heavy elements is 1% in the globular clusters versus 2–3% in the galactic clusters.

•PHOTOGRAPHIC CREDITS

The photographs that illustrate the text have been kindly furnished by

M. L. Bonelli Righini, Istituto e Museo di Storia della Scienza, Firenze, Italy (figure 69)

A Bruzek, Fraunhofer Institüt, Freiburg, German Federal Republic (figure 17)

Cerro Tololo Inter American Observatory, La Serena, Chile (plate 10) Dennis di Cicco of the magazine *Sky and Telescope*, Cambridge, Massachusetts (plate 6)

Gruppo Astrofili, Osservatorio San Vittore, Bologna, Italy (figures 3, 75, 90)

Hale Observatories, Pasadena, California (figures 2, 7, 9, 40, 58, 59, 60, 62, 67, 68, 80, 81, 91)

High Altitude Observatory of NCAR, Boulder, Colorado (figure 20)

Jet Propulsion Laboratory, NASA, Pasadena, California; furnished in part by M. Fulchignoni (figures 22, 23, 24, 25, 26, 27 and the figure on p. 6; plates 3, 4)

Kitt Peak National Observatory, Association of Universities for Research in Astronomy, Inc., Tucson, Arizona (figures 61, 72, 73, 79; plates 7, 8, 9, 11, 12)

Laboratorio di Radioastronomia del CNR, Bologna, Italy (figure 63)

Lick Observatory, University of California, Santa Cruz, California (figures 57, 64, 76, 77)

F. D. Miller, Department of Astronomy, University of Michigan, Michigan (plate 5)

David Moore (figure 1)

Mount Wilson and Palomar Observatories, Pasadena, California (figure 74)

Netherlands Foundation for Radio Astronomy, Netherlands (figure 88)

Osservatorio Astronomico di Capodimonte, Napoli, Italy (figures 4, 6, 8, 10, 11, 12, 13, 21, 28, 29, 30, 32, 33, 34, 35, 36, 37, 38, 39, 41, 42, 43, 44, 45, 46, 47, 48, 49, 50, 52, 53, 54, 55, 56, 64, 66, 70, 71, 82, 83, 84, 85, 86, 87, 89, 92, 93, 94, 95, drawn by S. Marcozzi; plate 1)

G. Righini, Osservatorio Astrofisico di Arcetri, Firenze, Italy (figures 16, 18)

J. Rösch, Observatoires du Pic du Midi et de Toulouse, France (figure 5)

L. Rosino, Osservatorio Astrofisico di Padova, Asiago, Italy (figures 31, 78)

Sacramento Peak Observatory, Association of Universities for Research in Astronomy, Inc., Sunspot, New Mexico (figures 14, 15, 19)

G. Vaiana, American Science and Engineering, Cambridge, Massachusetts (plate 2)

Yerkes Observatory, University of Chicago, Wisconsin (figure 51)

INDEX

Electrostatic attraction, 91
Elements
  in celestial objects, 112
  in massive stars, 254
  in stars, 119–120
Emission coefficient, 82, 85,
  86–87
Emission nebulae, 196, 197
Emissivity, stellar. *See* Stellar
  emissivity
Energy. *See also* Light; Photon of
  energy
  and mass, 230
  in a nova, 157
  solar, 16–20
Energy levels, of electrons, 93–96
Equilibrium, conditions of, 82
Eros, 72
Eta Carinae nebula, xxvi
Evolution of stars, 237–251
  H-R diagram and, 229–230
  model for, 227–228
  path of, 243, 244
  variation in size during, 248
Explosions. *See* Supernovae

Faculae, 35, 37, 39, 40
Flare stars, 37, 154, 155
Fluorescent emission, 180–181,
  196
Forbidden transition, 218
Force of attraction, and mass,
  126–127
Fraunhofer lines, 104
Free-bound transitions, 102
Free-free transitions, 102
Frequency of a wave, 83, 84
Fusion of hydrogen atoms
  in stars, 245
  in the sun, 19

Galactic clusters, 184, 202–204,
  207, 264. *See also* Clusters
  ages of, 263
  H-R diagrams for, 210
Galactic halo, 185
Galactic nucleus, stars in the, 215
Galaxies
  dwarf, 146

earth's overview of, 213–222
macroscopic view of, 181–184
masses of, 215–216
model for formation of, 263–265
photographs of, 186
rotation of, 216–217
Galileo, 60, 67, 127, 183
  Sidereus nuncius, 67
  telescopes of, 182
Gamma rays, 83, 84
Gas
  behavior of, 193
  density of, 102–103, 200
  excitation of atoms in, 95–96
  interstellar dust and, 185,
    188–202
  stellar (*see* Stellar gas)
Gas jets. *See* Spicules
Gas shell, 177
Giant convection cells, 27
Giant stars, 153, 177, 246–251
Globular clusters, 184, 203, 205,
  206, 264. *See also* Clusters
  ages of, 263
  H-R diagrams for, 208–210, 211
  measuring radial velocities of,
    215, 216
  M 3, 209
  number of, 213
Globules, 188, 238
Granulation, of photosphere, 24,
  25. *See also* Supergranulation
Gravitation, Newton's law of uni-
  versal, 67, 126, 128
Greenhouse effect, Venus and, 54

Halley's comet, 13
Harvard spectral classes, 116
Helium, 91, 196, 245
Helium flash, 247
Helium nuclei, 229, 232
Herschel, William, 127, 128, 183,
  184
Hertzsprung, E., 120
Hertzsprung-Russell diagram. *See*
  H-R diagram
HI regions, H atoms in, 201
HII (ionized hydrogen) regions,
  197, 200

Parallax
  calculation of, 124
  of a star, 70
  statistical, 147–151
Parsec, 70
Pascoli, Giovanni, 65
Pendulum, and analogy with
    electron energies, 91
Period-luminosity relation,
    146–147, 152
Phobos, 57, 59, 60
Photographing stars, 124
  in color, 88
  motion and, 148–149
  temperature and, 106
Photometer, as alternative to
    spectrum, 193
Photometric doubles, 133–134,
    135
Photon of energy, 91, 92, 95, 97
Photosphere, solar, 23–34, 104
  brightness of, 35
  convective motions in, 25
  granular structure of, 24
  sunspots and, 27–34
Physical laws of universe, 109
Physics, reformulations of, 17, 19
Pickering, E. C., 137
Planckian curves, 85, 86, 89, 107,
    110
Planck's law, 163
Planetary model of atoms, 90–91
Planetary nebulae, 177–181, 250
Planets, 44–63, 74–75. See also
    specific planets
  biological life on, 48
  chemical composition of, 45–46
  general characteristics of, 46–47
  masses of, 7–8
  movements of, 68
  number of, 10
  satellites of, 11
Pleiades, 194, 262
  H-R diagrams for, 210, 211
Pleiades cluster, 203
Pluto, 7
  general data on, 75
Pogson's relation, 77, 111
Polaris, magnitude of, 81

Polarized light, 169–170
Population II stars. See Globular
    clusters
Praesepe cluster, 203
Prenovae, 156
Prism spectrograph, 99, 100, 101
Prism technique, 109
Prominences, solar, 37, 39, 41,
    42, 43
Proper motions, 147–151
Proton-proton reaction, 232
Protons, 22, 90, 91, 201, 229
Protostar
  contraction of, 242–243, 245
  evolution to star of, 241–246
  formation of, 238
  H-R diagram and, 242
Proxima Centauri, 70
Pulsars, 171–175, 255–256. See
    also Neutron stars
  sizes of, 175, 176
  and supernovae, 176
Pulsating star, 144, 146

Quantum of light, 91

Radial velocity, 115–116, 136,
    137
  measuring, 149
Radiation. See also Blackbody;
    Energy; Solar energy
  abnormal flux of (see Flare stars)
  atomic, 91
  emitted by stars, 97–98
  infrared (see Infrared (ir)
    radiation)
  kinds of, 83, 84
  origin of, 89–98
  solar emission of, xviii
  synchrotron, 169, 171, 200, 255
  21-cm, 200–202, 218–219
  ultraviolet (see Ultraviolet (uv)
    radiation)
Radiation absorption. See
    Blackbody
Radio emission
  nebulae and, 163, 165, 167, 169
  21-cm, 218, 219

Radio-frequency emission, and
interstellar gas, 196
Radio source, pulsating. *See*
Pulsars
Radiotelescopes, 168, 171, 173,
202
Radio waves, 83, 200
angular, 130
R Coronae Borealis, 154, 155
Red giants, 177
evolution to black dwarfs,
250–251
formation of, 246–249
Red supergiants, 153
Reflection nebulosity, 193–195
Riccioli, Giovanni Battista, 127
Rosette nebula, 238, 239, 240
Rotational velocity of the sun, 27,
28
RR Lyrae stars, 151, 152, 153,
209
RR Lyrae variables, 215
Russell, H. N., 120
Russell-Vogt theorem, 236, 237

Sagittarius
Lagoon nebula in, xxii
Omega nebula in, xxv
star field in, 191
Trifid nebula in, xxiv
Satellites of planets, 11
Saturn, 7
general data on, 75
Savary, F., 128
Schwarzschild limit, 256, 257
Science
defining, 18–19
military and, 63–64
social relevance of, 5
theoretical nature of, 18
Scientific method, 118–119, 167
meaning of hypothesis in, 90
Scientists
as ordinary people, 122
role of, 18
Secchi, Angelo, 34
Serpens, 190
Aquilla nebula in, xxvii

Shapley, Harlow, 147
Shell stars, 177
Sirius, 81, 138
61 Cygni, 70
Sky, blue color of, 54–55
Solar activity, cycle of, 30, 31, 32,
34, 35
Solar apex, 216
Solar chromosphere. *See*
Chromosphere
Solar corona. *See* Corona
Solar diameter. *See* Photosphere
Solar energy, 16–20
Solar granulation, 25
Solar luminosity, 110, 230
Solar magnitude, 15–16, 21, 47,
81
Solar mass, 230–231, 234–235
Solar photosphere. *See*
Photosphere
Solar prominences. *See*
Prominences
Solar spectrum, 104, 105,
195–196
Solar system, 10–15
model, 7
space occupied by, 7–8
Solar temperature, 232
Solar wind, 12, 22–23
Sound waves, 113
Space, 7–8, 12
Spacecraft, 53. *See also Mariner;
Venera; Viking*
landing on Mars, 62
Space exploration
and the planets, 48–63
social and political value of,
63–64
Space velocity, 149, 151, 216
Spectral classes of stars, 116–120
extended atmospheres and, 177
H-R diagrams and, 123, 124,
125
surface temperatures and, 208
Spectral lines, 100
gas density and, 102–103
novae and, 157
Spectrograph, 99, 100, 109

Spectroheliogram, 38, 40
Spectroscope, 99, 103
Spectroscopic binaries, 134,
   136–138
Spectroscopic parallaxes, 124
Spectrum
   of blackbody, 99, 100
   electromagnetic, 83, 84
   of a light source, 98
   line, 112
   stellar, 93, 98–104, 124
   of sun, 104, 105
   variations in (*see* Spectroscopic
      binaries)
Speed of light, 7, 83, 98
Spicules, 34, 35, 36
Spiral arms, of gas and interstellar dust, 185, 209
Spiral galaxies, 213
Star associations, 207–208, 209
Star formation, 237–251
   from main sequence to red
      giants, 246–249
   from protostar to star, 241–246
   from red giant to black dwarf,
      250–251
Stars
   atmospheres of (*see* Stellar
      atmospheres)
   associations of (*see* Associations
      of stars)
   binary (*see* Binary systems)
   as blackbodies (*see* Blackbody,
      star as)
   vs. blackbodies, 107–108
   black dwarf (*see* Black dwarfs)
   blue (*see* Blue stars)
   blue giant (*see* Blue giants)
   brightness of (*see* Brightness of
      stars; Luminosity; Stellar
      magnitudes)
   circumpolar, xxii
   classification of, 116–120
   clusters and associations of,
      202–208
   colors and temperatures of (*see*
      Colors and temperatures of
      stars)

definition of, 45
distances of (*see* Stellar
   distances)
double (*see* Double stars)
dwarf (*see* Dwarf stars)
elements in (*see* Elements, in
   stars)
energy of (*see* Stellar energy)
evolution of (*see* Stellar
   evolution)
exploding (*see* Supernovae)
flare (*see* Flare stars)
giant (*see* Giant stars)
high-velocity, 217, 218
hot (*see* Hot stars)
interiors of, 232–237
luminosity of (*see* Luminosity)
magnitudes of (*see* Stellar
   magnitudes)
masses of, 126–127, 128, 132
massive, 251–256
measurements of, 70
models of, 233–236
motions of, 68, 69, 113–116,
   147–151
neutron (*see* Neutron stars)
parallaxes of, 70, 124, 147–151
populations of (*see* Stellar
   populations)
pulsating (*see* Pulsars)
radiation of, 97–98 (*see also*
   Stellar spectra)
red giant (*see* Red giants)
red supergiant (*see* Red
   supergiants)
shell (*see* Shell stars)
shooting (*see* Meteor craters)
sizes of (*see* Stellar magnitudes)
small, 176 (*see also* Pulsars)
spectra of, 98–104, 116–120
speeds of (*see* Doppler effect)
supergiant (*see* Supergiants)
temperatures of, 81–82, 89,
   106–109, 119
variable (*see* Variable stars)
white dwarf (*see* White dwarfs)
Statistical parallaxes, 147–151
Stefan's law, 110

Stellar atmospheres
  chemical analysis of, 100
  extended, 177–181
  model of, 104–106
  temperatures of, 108
Stellar distances, 60, 67–72, 124.
  See also Proper motions
Stellar emissivity, 110
Stellar energy
  distribution of (*see* Color
    temperature)
  output of, 110
  source of, 19, 20, 229–232
  total flux of (*see* Effective
    temperature)
Stellar evolution, 237–251
  H-R diagram and, 228–229, 230
  model for, 227–228
Stellar gas, 97
Stellar magnitudes, 73–81, 88,
  110–111
Stellar populations, 208, 213
Stellar spectra, 98–104
  classification of, 116–120
Sun
  energy of (*see* Solar energy)
  external layers of (*see* Chromo-
    sphere; Corona; Photosphere)
  formation of, 245–246
  general data on, 74
  luminosity of (*see* Solar
    luminosity)
  mass of (*see* Solar mass)
  motions of, 27, 28, 216
  planets of, 74–75
  temperature of (*see* Solar
    temperature)
Sunlight, scattering of, 54–55
Sunspots, 26, 27–34, 41
  chromospheric region around,
    39
  cycles of, 30–34
  and faculae, 37
  formation of, 30
  magnetism and, 30, 31, 32, 34
  temperature and size of, 27
Supergiants, 123, 141, 153
  density of, 140
  type A, 157

Supergranulation, 24, 27
Supernovae, 154, 155, 158,
  160–169
  formation of, 253–254
  gaseous remnant of (*see* Crab
    nebula; Veil nebula)
  light curve of, 160
  pulsars and, 176
Surface temperatures
  of stars, 89
  of sun, 232
Swift, Jonathan, 57
Synchrotron radiation, 169, 171,
  200, 255

Tangential velocity, 150
Taurus, 160
T Tauri stars, 155, 242
Telescopes, 70, 76–77, 78. *See*
  also Radiotelescopes
  Galileo's, 182
Temperature index, and color in-
  dex, 88
Temperatures of stars, 89,
  106–109, 119. *See also* H-R
  diagram
  absolute magnitude and, 123
  central, 232
  radiation and, 85, 86
  surface, 81–82
Theodolite, 69
Transitions, 95, 96, 218
Trifid nebula, xxiv, 191

U Geminorium stars, 154
Ultraviolet (uv) radiation, 83, 84
Unicorn, constellation, 238
Universe, beginning of, 253
Uranus, 7
  general data on, 75
Ursa Major, 127, 128, 148
UV Ceti type, 155

Van de Hulst, H. C., 202
Van Maanen's star, 138
Variable stars, 118–119, 142–169
  irregular, 153–169
  regular, 143–147, 151–153

separation of, 118–119
Vega, 88, 216
  color of, 87
  magnitude of, 81
Veil nebula, 162, 163
Velocity
  low, of stars, 114
  radial, 115–116, 136, 137, 149
  rotational, solar, 27, 28
  space, 149, 151, 216
  tangential, 149, 150
Velocity curve, 137
Venera, 53, 54
Venus, xix, 7, 53–55, 56
  general data on, 74
  phases of, 49
  radar map of, 56
Viking, 62
Visible spectrum, 83, 84
Visual binaries, 127–133
Visual magnitude, apparent, 88
Vogel, H., 134

Wavelength
  radiation and, 82–83, 85, 86
  radio observations and, 218
  sound and, 113
  spectrographs and, 100
White dwarfs, 122, 138
  formation of, 250–251
  massive stars and, 254
Wright, T., 183
W Virginis stars, 152

X rays, 83, 84
  nebulae and, 171
  sources of, 176

Z Camelopardalis stars, 154
Zodiacal light, 10, 12, 21